CASTE AND ECOLOGY

IN THE

Social Insects

MONOGRAPHS IN POPULATION BIOLOGY

EDITED BY ROBERT M. MAY

CASTE AND ECOLOGY

IN THE

Social Insects

GEORGE F. OSTER AND

EDWARD O. WILSON

PRINCETON, NEW JERSEY

PRINCETON UNIVERSITY PRESS

1978

Copyright © 1978 by Princeton University Press
Published by Princeton University Press, Princeton, New Jersey
In the United Kingdom: Princeton University Press, Guildford,
Surrey
ALL RIGHTS RESERVED
Library of Congress Cataloging in Publication Data will be
found on the last printed page of this book
This book has been composed in Monophoto Baskerville
Clothbound editions of Princeton University Press books are
printed on acid-free paper, and binding materials are chosen for
strength and durability.
Printed in the United States of America by Princeton
University Press, Princeton, New Jersey

Aharon Katzir-Katchalsky

Preface

This monograph is not intended to be a finished treatise on the theory of castes in social insects. Rather, it is a prolegomenon—a beginning effort to make some sense out of the veritable mountain of empirical findings gathered by hundreds of investigators over decades of research. By focusing attention on the ecological and evolutionary aspects of caste, as distinct from developmental and physiological processes, we hope to provide the beginnings of a unifying theoretical framework.

When we first conceived the idea of developing a "theory of caste" the task seemed manageable. Here was a neat, circumscribed evolutionary puzzle that had troubled biologists since Darwin. Moreover, the empirical data, while voluminous, seemed to hint at an underlying simplicity. Perhaps, with a bit of luck and cleverness, we could find the "answer"—a single new way of looking at the problem that would disclose the raison d'être of caste polymorphism in social insects and cause all the data to fall into place. This conceit was shortlived. The deeper we delved the more we came to see the phenomenon of caste structure as a microcosm for much larger evolutionary issues in sociobiology.

The most important realization that changed our view was the observation that caste polymorphism represents but one evolutionary strategy to enhance ergonomic efficiency. An equally potent strategy is "behavioral flexibility." Indeed, once a species adopts sufficiently complex individual behavior that can respond to fluctuations in the environment, selective forces promoting physical polymorphism are largely neutralized. Thus, of all the 263 living ant genera, only 44 contain species that display any prominent degree of polymorphism within the worker caste; the others clearly have been able to

vii

do without it. The ultimate reason why a species takes the evolutionary road to physical polymorphism may be imponderable. Indeed, current evolutionary thought gives a much larger role to historical accident than in the past. The function of science is to explain what can be explained, not what should be explained.

Therefore, the most that can be accomplished from evolutionary retrospection might be to characterize the agents of natural selection that promote one or the other of multiple strategies. The best way of doing this, in our view, is to construct quantitative models. The reason quantification is necessary is the ambitendent nature of the evolutionary process. As we shall stress repeatedly throughout this book, the direction of evolutionary change is the result of conflicting adaptive requirements. The optimal adaptation is almost always a compromise between selective forces whose magnitudes are commensurable. Thus, evolutionary problems can seldom be resolved by verbal arguments alone, except in exceptional instances where there is a single overwhelming selective force. In most cases, it is not a question of which forces operate, since their effects are manifest, but rather of their relative magnitudes. The evolutionary solution reached by a species is a quantitative rather than a qualitative problem. For this reason we have chosen to organize the entire book around a series of mathematical models, which are conceived to have two purposes: (1) to provide an unambiguous conceptual structure for the theory, and (2) to serve as a guide for future empirical research.

Like most biologists concerned with theory, we appreciate that models do not themselves provide definitive answers to biological questions. Mathematics is, after all, little more than scaffolding upon which all manner of facts about the real world can be arranged, hopefully into a more esthetic and useful structure. Mathematical theory is the science of all conceivable worlds. The most important role models play in science is to help us to perceive a problem more clearly and to generate thoughts that might not otherwise have occurred.

From one point of view, then, our search for a general theory of caste was a failure: we broke no hidden code. But, as it so often happens in science, the effort of the search was rewarding and the disillusionment enlightening. We hope that others will be stimulated to view the problem from different perspectives or to carry out some of the empirical investigations that our models have suggested.

A word is in order concerning the basic approach we have selected. We have chosen to view the evolution and ecology of insect castes from the perspective of ergonomic efficiency. This appears to be the only theme that is both unifying and sufficiently explicit to offer some hope of empirical verification. Consequently, most of the models presented in this monograph are of an economic nature. The question is repeatedly asked: What is the most efficient way of allocating resources among available alternatives? Optimization theory plays the central role in our treatment of the caste problem, in keeping with a long tradition in ecology and evolutionary theory. Recently, however, several authors have voiced important criticisms against the uncritical use of optimization arguments (Sahlins, 1976; Levins, 1977; Lewontin, 1977a,b). We, too, have been acutely aware of these shortcomings; and so, if only to counterbalance the force of the preceding chapters, we have set down in Chapter Eight our own views on the limitations of optimality arguments in evolutionary biology. Such considerations, nevertheless, do not vitiate the guiding proposition that natural selection generally acts to increase ergonomic efficiency.

The chapters are organized in a sequence roughly corresponding to descending levels in the units of natural selection. In Chapter Two, we examine the general life cycle patterns of social insect colonies. The models there deal with the colony itself as the unit of selection.

Chapter Three demonstrates how selection for ergonomic efficiency is related to the maximization of the inclusive fitness of the individuals in the colony (queens, workers, and drones).

Chapter Four contains a review of the general biological properties of caste systems.

In Chapter Five we introduce the definitions of the caste and resource distribution functions, and examine the selective forces limiting caste specialization.

Chapter Six consists of an analysis of the countervailing selective forces that promote caste proliferation. The necessity of coping with an uncertain environment plays a crucial role in this regard, and so we introduce the important notion of tradeoffs between risk and return in colony ergonomics. The general role of caste efficiency is then analyzed as a determinant of caste distributions.

If ergonomic efficiency is preeminent in shaping caste evolution, the role of foraging efficiency on colony ergonomics must be examined in detail. There is a substantial literature on optimal foraging theory; Chapter Seven offers explicit suggestions on how to adopt these theories to the foraging strategies of social insects.

In Chapter Eight we face the unstated assumptions underlying all of the optimization models developed in the preceding chapters and present our views on the proper role of optimization models in evolutionary biology.

In spite of a large backlog of empirical information, few of the issues addressed in the course of the modeling efforts have been definitively resolved. Therefore, in Chapter Nine, we describe our own perspectives on the outstanding issues and offer a "shopping list" of empirical studies needed to settle these questions.

The problem of how to integrate mathematical reasoning with biology is one that has no unique solution. In principle, we would have liked the book to be equally accessible to both mathematical and empirical biologists. But we suspect that both camps will not be totally satisfied. Our compromise has been to keep the mathematics to a minimum within the body of the chapters, while relegating the mathematical details to appendixes—a procedure previously used with success by Robert May in his contribution to this series. A complete mathematical treatment of all of the models would have necessitated appendixes more voluminous than the book itself. And so we have restricted ourselves in most instances to a bare sketch of the mathematical details, referring the reader to the cited literature for whatever amplification is required.

In all but a few cases the empirical studies required to test the models have not yet been carried out. Often, we have done little more than formulate a model, in the hope that it will at least stimulate others either to pursue the analysis further or to suggest an entrée to empirical research that can confirm or refute the theory.

A large portion of an article by Hölldobler and Wilson (1977a) and several paragraphs of articles by Oster (1976) and Macevicz and Oster (1976) have been repeated with little change in Chapters Two and Seven of this book; we thank the publishers of *The American Naturalist*, *Naturwissenschaften* and *Behavioral Ecology and Sociobiology* for permission to use this material. The following figures have also been reproduced with permission: Figures 8.4, 8.6, and 8.8 from *The Insect Societies*, and Figures 2.2 and 2.5 from *Sociobiology: The New Synthesis*, both by E. O. Wilson (Belknap Press of Harvard University Press); and Figures 1 and 3 from E. O. Wilson, "Behavioral discretization and the number of castes in an ant species," *Behavioral Ecology and Sociobiology*, 1(2): 141–154 (1976).

Finally, we wish to express here our gratitude to the numerous colleagues whose judgment and intellectual stimulation are represented on virtually every page of this monograph. A partial list includes: Tracy Allen, Gary Alpert, David Auslander, Frank Benford, William L. Brown, Arthur Caplan,

Dan Cohen, Joel E. Cohen, Robert K. Colwell, Howell Daly, Ilan Eshel, Adrian Forsyth, Michael E. Gilpin, John Guckenheimer, Bernd Heinrich, Bert Hölldobler, Jan Kwiatkowski, George Leitmann, Simon A. Levin, Richard C. Lewontin, Steven Macevicz, Robert M. May, John Maynard Smith, Majdedin Mirmirani, Alan Perelson, Chris Plowright, Sol Rocklin, Jonathan Roughgarden, Jon Seger, Steven Stearns, Yasundo Takahashi, Robert L. Trivers, and Arthur Winfree. Kathleen M. Horton assisted in bibliographic research, typed the manuscript, and provided invaluable editorial expertise. The authors were able to work closely together while holding fellowships from the John Simon Guggenheim Foundation.

George F. Oster
Berkeley, California

Edward O. Wilson
Cambridge, Massachusetts

September 16, 1977

Contents

CONTENTS

CASTE AND ECOLOGY

IN THE

Social Insects

FRONTISPIECE. The caste system of the African weaver ant *Oecophylla longinoda*. The large queen is surrounded by major workers, who groom her, feed her by regurgitation, and receive eggs as she lays them from the tip of her abdomen. To the right side of her head a major worker doubles forward to lay a trophic egg, which will be fed to the queen. Major workers also attend the later developmental stages of the brood; to the right of the queen a major worker regurgitates liquid food to a mature larva, while to her left another individual carries a pupa. In the foreground three minor workers attend eggs and small larvae, the task on which they are specialized. (Original illustration by Turid Hölldobler.)

CHAPTER ONE

Why Is Caste Important?

Caste and division of labor lie at the heart of colonial organization in the social insects. What makes an ant colony distinct from a cluster of butterflies, a swarm of midges, or for that matter a flock of birds, is its internal organization—the differentiation of its members into castes, the division of labor based on caste, and the coordination and integration of the activities that generate an overall pattern of behavior beyond the reach of a simple aggregation of individuals.

Indeed, caste is one of the essential criteria of eusociality. All of the more than 10,000 ant species, 2,200 termite species, the thousand or so social wasps, and the several thousand species of social bees have the following three traits in common: (1) cooperation in caring for the young; (2) overlap of at least two generations capable of contributing to colony labor; and (3) reproductive division of labor, with a mostly or wholly sterile worker caste working on behalf of individuals engaged in reproduction. Of these, division of labor is the single trait essential to the advanced modes of colonial existence.

One of the puzzles of sociobiology is that, despite the fact that caste is central to the life of the social insects, it has evolved in notably irregular patterns and often only to a limited degree. Systems of physical subcastes—the differentiation of the workers into minors, soldiers, and other forms—are virtually absent in the bees and wasps. As shown in Table 1.1 and Figure 1.1, these systems are also limited to a minority of the ant genera, and the most elaborate systems, possessed by *Eciton*, *Atta*, *Daceton*, and *Pheidologeton*, still contain no more than three truly distinct forms. The termites have more complex caste systems than ants and other social hymenopterans, but at least

3

WHY IS CASTE IMPORTANT?

TABLE 1.1. All of the ant genera known in 1973, listed according to whether the species lack a worker caste, or all the species are monomorphic, or at least some of the species are polymorphic enough to have two or three easily distinguishable worker subcastes. The generic names are taken from the checklist by W. L. Brown (1973) and presented in the conventional taxonomic order, which is based on inferred phylogenetic relationships.

NO WORKER SUBCASTES (WORKERLESS SOCIAL PARASITES)

MYRMICINAE: *Anergates, Teleutomyrmex, Hagioxenus*

ALL SPECIES IN THE GENUS ARE MONOMORPHIC (ONE WORKER SUBCASTE ONLY)

MYRMECIINAE: *Nothomyrmecia.*
PONERINAE: *Amblyopone, Mystrium, Myopopone, Prionopelta, Onychomyrmex, Apomyrma, Paraponera, Acanthoponera, Heteroponera, Rhytidoponera, Ectatomma, Aulacopone, Gnamptogenys, Proceratium, Discothyrea, Typhlomyrmex, Platythyrea, Probolomyrmex, Sphinctomyrmex, Cerapachys, Leptanilloides, Simopone, Cylindromyrmex, Acanthostichus, Thaumatomyrmex, Harpegnathos, Diacamma, Centromyrmex, Dinoponera, Streblognathus, Paltothyreus, Odontoponera, Pachycondyla, Ophthalmopone, Hagensia, Euponera, Brachyponera, Cryptopone, Simopelta, Belonopelta, Emeryopone, Ponera, Hypoponera, Plectroctena, Psalidomyrmex, Asphinctopone, Leptogenys, Prionogenys, Odontomachus.*
ECITONINAE: *Nomamyrmex, Neivamyrmex.*
LEPTANILLINAE: *Leptanilla, Leptomesites.*
DORYLINAE: *Aenictus.*
PSEUDOMYRMECINAE: *Pseudomyrmex.*
MYRMICINAE: *Myrmica, Manica, Hylomyrma, Ephebomyrmex, Aphaenogaster, Goniomma, Oxyopomyrmex, Proatta, Stenamma, Rogeria, Lordomyrma, Lachnomyrmex, Geognomicus, Dacetinops, Adelomyrmex, Prodicroaspis, Promeranoplus, Calyptomyrmex, Mayriella, Meranoplus, Podomyrma, Dilobocondyla, Terataner, Atopomyrmex, Poecilomyrma, Atopula, Brunella, Ireneopone, Peronomyrmex, Vollenhovia, Rhopalomastix, Metapone, Melissotarsus, Liomyrmex, Leptothorax, Harpagoxenus, Tetramorium, Xiphomyrmex, Decamorium, Rhoptromyrmex, Triglyphothrix, Eutetramorium, Strongylognathus, Macromischoides, Tetramyrma, Huberia, Chelaner, Syllophopsis, Anillomyrma, Diplomorium, Paedalgus, Allomerus, Megalomyrmex, Nothidris, Oxyepoecus, Carebara, Carebarella, Tranopelta, Trigonogaster, Lophomyrmex, Stereomyrmex, Xenomyrmex, Myrmecina, Pristomyrmex, Perissomyrmex, Ocymyrmex, Myrmicaria, Cardiocondyla, Ochetomyrmex, Romblonella, Willowsiella, Stegomyrmex, Phalacromyrmex, Tatuidris, Basiceros, Aspididris, Creightonidris, Octostruma, Rhopalothrix, Eurhopalothrix, Cataulacus, Acanthognathus, Epopostruma, Mesostruma, Trichoscapa, Colobostruma, Microdaceton, Neostruma, Smithistruma, Kyidris, Serrastruma, Glamyromyrmex,*

4

Table 1.1 (*continued*)

Dorisidris, Dysedrognathus, Epitritus, Pentastruma, Miccostruma, Quadristruma, Tingimyrmex, Procryptocerus, Apterostigma, Cyphomyrmex, Mycocepurus, Myrmicocrypta, Mycetarotes, Trachymyrmex, Sericomyrmex.
DOLICHODERINAE: *Leptomyrmex, Dolichoderus, Monoceratoclinea, Semonius, Axinidris, Turneria, Froggattella, Dorymyrmex, Forelius, Neoforelius, Bothriomyrmex, Engramma, Tapinoma, Ecphorella, Technomyrmex, Anillidris.*
FORMICINAE: *Myrmoteras, Notoncus, Pseudonotoncus, Prolasius, Lasiophanes, Acropyga, Aphomomyrmex, Cladomyrma, Brachymyrmex, Myrmelachista, Pseudaphomomyrmex, Plagiolepis, Anoplolepis, Acantholepis, Stigmacros, Prenolepis, Paratrechina, Lasius, Acanthomyops, Teratomyrmex, Polyergus, Rossomyrmex, Gigantiops, Santschiella, Opisthopsis, Overbeckia, Dendromyrmex, Calomyrmex, Echinopla, Polyrhachis.*

AT LEAST ONE SPECIES POLYMORPHIC, POSSESSING TWO WORKER SUBCASTES

MYRMECIINAE: *Myrmecia.*
PONERINAE: *Megaponera.*
ECITONINAE: *Labidus, Cheliomyrmex.*
DORYLINAE: *Dorylus.*
PSEUDOMYRMECINAE: *Tetraponera.*
MYRMICINAE: *Pogonomyrmex, Messor, Veromessor, Pheidole, Monomorium, Solenopsis, Oligomyrmex, Adlerzia, Machomyrma, Anisopheidole, Acanthomyrmex, Crematogaster, Orectognathus, Strumigenys, Cephalotes, Zacryptocerus, Acromyrmex.*
DOLICHODERINAE: *Aneuretus, Liometopum, Iridomyrmex, Azteca, Zatapinoma.*
FORMICINAE: *Oecophylla, Gesomyrmex, Myrmecorhynchus, Melophorus, Euprenolepis, Pseudolasius, Myrmecocystus, Cataglyphis, Proformica, Formica, Notostigma, Camponotus.*

AT LEAST ONE SPECIES POLYMORPHIC, POSSESSING THREE SUBCASTES

ECITONINAE: *Eciton.*
MYRMICINAE: *Pheidologeton, Daceton, Atta.*

DEGREE OF POLYMORPHISM UNKNOWN

LEPTANILLINAE: *Phaulomyrma, Scyphodon, Noonilla.*
DORYLINAE: *Aenictogiton.*
DOLICHODERINAE: *Linepithema.*
FORMICINAE: *Phasmomyrmex, Forelophilus.*

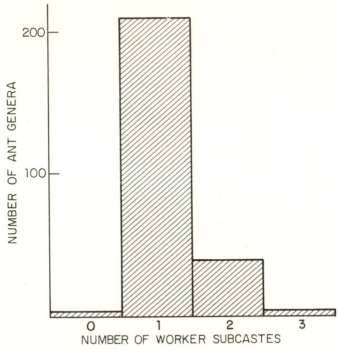

FIGURE 1.1. The frequency of ant genera that are (1) workerless social parasites (possessing no worker subcastes), (2) exclusively mono-morphic (possessing one worker subcaste), or (3) containing at least one polymorphic species (which possess two or three worker subcastes).

one genus (*Anoplotermes*) has monomorphic species. This great variety in caste, and the division of labor based upon it, provides a major challenge to evolutionary theory.

1.1. THE SIMPLIFICATION OF INDIVIDUAL BEHAVIOR

One fact about insect social life of immediate importance is that the behavior of each individual is greatly simplified com-pared to the activities of, say, a solitary wasp. Division of labor

6

is accompanied by a loss in the total behavioral repertory of individual caste members. For example, a solitary female wasp belonging to a phylogenetically advanced group such as the Eumenidae, Pompilidae, or Sphecidae performs an impressive array of behaviors during the course of her life: in a typical sequence she mates, locates a suitable nest site, constructs an elaborate nest, locates the correct prey, stings and paralyzes the prey, returns to the nest, lays an egg on the prey, and seals the nest. Many ant queens are similarly "totipotent" by hyme-nopterous standards: they leave the mother nest, mate, locate a suitable nest site, construct a relatively simple nest, and rear a brood of workers. As soon as the first workers emerge from the pupae and commence their labors, the queen drastically reduces her repertory. For the rest of her life she does little more than receive food and lay eggs. The workers fill in the remaining tasks, but they never engage in a nuptial flight and only rarely lay eggs. As the colony grows in size and additional physical castes and age groups are added, the repertory of individuals diminishes still more. Some workers might display an idio-syncratic inclination to attend brood and perhaps later to seek honeydew, but undertake little else. Others might largely bypass brood care to mend the nest and forage. Major workers are still more rigidly restricted; they typically perform only a single task such as defense, food storage, or the milling of seeds. The important point is that the colony as a whole is behaviorally equivalent to a totipotent solitary individual but possesses a much higher ergonomic efficiency.

As the division of labor has increased in the course of evolu-tion across species, the average repertory size of each worker has decreased, while the total repertory size of the colony as a whole has typically increased only slightly over that of the totipotent queen. If we were to imagine the lifetime repertory of an ant queen as a set of 8 behaviors that can be referred to arbitrarily as A through H, with the mating activities designated

A and *B*, the repertory of a fully grown ant colony might be *C-K*, with the lifetime repertories of workers belonging to various castes represented respectively as *C-H*, *F-G*, *I-K*, and so forth. Another way of viewing the matter is by minimum specification, illustrated in Figure 1.2. This criterion defines the complexity of a system as the number of constituent units that need to be enumerated in order fully to characterize the system. Fewer solitary wasps are required to encompass the full behavioral repertory of their species than is the case of a typical ant species. As the repertories of more individual ants are added, the total colony repertory eventually climbs above that of the wasps—but not by very much. The total worker repertories have been estimated for several diverse species: *Leptothorax*

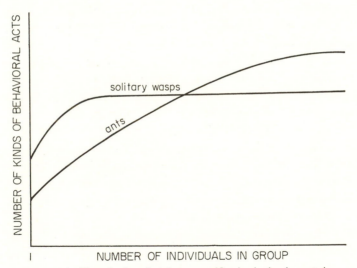

FIGURE 1.2. The criterion of minimum specification in the characterization of social complexity in two species of insects. One solitary wasp has a larger behavioral repertory than one ant, and a smaller group of wasps is required for the display of the entire species repertory. As the ant group is enlarged, the full repertory is encompassed more slowly. It eventually becomes larger but not by a very great margin. These qualitative statements are believed to be correct, but the details of the curves are imaginary. (From Wilson, 1975a.)

curvispinosus, 29 (Wilson, 1975b); *Pheidole dentata*, 26 (Wilson, 1976a); *Solenopsis geminata* and *S. invicta*, 20 (Wilson, 1978); *Cephalotes atratus*, 40 (Corn, 1976); *Zacryptocerus varians*, 42 (Wilson, 1976b). While these estimates are subject to some variation according to the system of behavioral classification used by the observer, we believe that they would only slightly exceed in size the repertories of many female solitary wasps, and they might even be smaller. When the sexual and nest-founding repertories of the ant queens are added, the total colony repertories almost certainly exceed those of the wasps—but still by a modest amount.

1.2. THE LACK OF GREAT BEHAVIORAL NOVELTY

Thus, contrary to popular belief, social life in the insects has not been built upon major increases in the complexity of behavior. Nor has it led to genuine breakthroughs in adaptations to the environments. Some entomologists, for example, consider the fungus-growing ants to be the pinnacle of social evolution. These insects live in an agricultural society, cultivating symbiotic fungi on fragments of caterpillar dung, vegetable detritus, or fresh leaves gathered from the area surrounding the nest. But a variety of solitary insects also cultivate fungi. Certain scolytid beetles, notably members of the genera *Gnathotrichus*, *Monarthrum*, and *Xyleborus*, rear their young in "cradles" carved in dead wood and provisioned at regular intervals with pellets of freshly gathered fungus mycelia. Their behavior rivals that of the fungus-growing ants in sophistication. Keeping aphids and other honeydew-producing homopterous insects is also a practice of certain ant species, which have therefore been called the "pastoralists" of the insect world. Yet solitary beetles of the silvanid genera *Coccidotrophus* and *Eunausibius* also milk mealybugs within cavities constructed in living plant petioles. After consuming the protein-rich pith of the petioles, the beetles

9

subsist entirely on the mealybug honeydew. In one of his more memorable passages, William Morton Wheeler referred to the army ants as the Huns and Tartars of the insect world. Their huge colonies (containing as many as 22 million workers in the case of *Dorylus wilverthi*) sweep large areas clean of insects and other small animal prey. Yet, once again, their predatory habits are different more in degree than in kind from the thousands of species of solitary predatory insects with which they compete. In short, the basis of the enormous ecological success of the social insects is not innovation in their methods of extracting energy from the environment, but rather in the scale of their operations.

Nor have social insects by and large produced major advances in defensive behavior. It is true that many have evolved alarm and recruitment systems that organize massive resistance to enemies, and special soldier castes are commonplace among the species of ants and termites. These advances are impressive, but no more so than the remarkable array of technically sophisticated chemical and mechanical defenses evolved by solitary insects (see, for example, the review by Eisner, 1970).

1.3. CONCURRENT OPERATIONS AS THE KEY ADVANCE

What makes a colony of social insects truly distinctive is its capacity to conduct all of its operations concurrently instead of sequentially. A colony-founding ant queen, like a solitary wasp female, can only complete one task at a time. In order to conduct the multiple functions necessary for survival, she must switch from one task to the next, undertaking a change from one very different set of stimuli to another. Within the nest and in the surrounding foraging ground, only one contingency can be met at a time, only a single opportunity seized. The well-developed colony, in contrast, blankets the area and meets most or all of the important contingencies as they appear—the arrival of an enemy or prey on the territory, the hunger signal of a larva, the collapse of a nest gallery, and so on through twenty or more such categories. As a consequence, the responses of a colony, in contrast to those of a solitary individual, are more massive, prompt, and thorough. They are also more efficient, because the specialized caste status of each colony member tends to match the contingency to which it responds. And finally they are more secure, because if one colony member fails in its attempt to complete a task another is likely to succeed. In the design of control devices, systems engineers commonly utilize parallel operations instead of series operations, so that the breakdown of one unit will not cause the failure of the whole device. It might be said that an insect colony—viewed as a "superorganism"—differs from an individual in that it is based on parallel operations and hence is less dependent on the precise functioning of all of its parts.

The option of parallel operation as opposed to series operation might have been crucial in the evolution of insect eusociality. When this proposition is expanded in a more formal manner, some unexpected results emerge. Consider a solitary wasp or colony-founding queen, which must perform all of the requisite behavioral acts in a series; failure to perform an act at the appropriate time leads to failure of the whole reproductive enterprise. If n distinct acts are required for success, and if p_i is the probability of performing the ith task correctly, then the probability of overall success (P) is

$$P(\text{success}) = \prod_{i=1}^{n} p_i. \qquad \text{(series)}$$

This is merely the probability that all acts are performed correctly. (See Fig. 1.3.) Thus, if a founding queen attempts to seal the nest, and fails to do so properly, all previous and subsequent acts might be wasted. The earlier excavation of the brood chamber, for example, might come to nought because a predator is able to enter it and consume all of the queen's subsequently laid eggs.

11

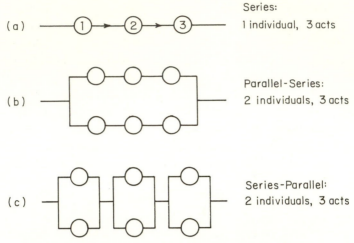

FIGURE 1.3. A comparison of the reliability of behavioral sequences. If $p_{ij} = p = 0.2$, then

$$P\text{ (series)} = (0.2)^3 = 0.008$$
$$P\text{ (parallel-series)} = 1 - [1 - (0.2)^3]^2 = 0.016$$
$$P\text{ (series-parallel)} = [1 - (1 - 0.2)^2]^3 = 0.047$$

In general, the reliability of (*c*) is greater than that of (*b*), which is, in turn, greater than that of (*a*).

However, if a group of individuals work together, the probability of overall success is always higher, except in the trivial cases of total individual competence ($p_i = 1$) or incompetence ($p_i = 0$). Thus, if there are m individuals, acting more or less independently, the probability that at least one individual will perform the required behavioral sequence correctly is:

$$P\text{(success)} = 1 - \prod_{j=1}^{m} \left(1 - \prod_{i=1}^{n} p_{ij}\right), \quad \text{(Parallel-series)}$$

where p_{ij} is the probability that the jth individual will perform the ith act successfully. This is clearly a conservative estimate since each individual is required to perform the complete task sequence separately. More realistically, each individual may complete an act or not, and it matters only that the correct

12

sequence be performed by any combination of individuals. The reliability of the operation is then:

$$P(\text{success}) = \prod_{i=1}^{n} \left[1 - \prod_{j=1}^{m} (1 - p_{ij}) \right] \quad \text{(Series-parallel)}$$

A comparison of these behavioral sequences is shown in Figure 1.3 for $p_{ij} = p$, all i and j. Finally, the nature of most behavioral acts involved in colony activities (for more precise definitions of these concepts, see Chapter Four) are not so rigidly organized that all out of a sequence of n acts must be performed successfully. More generally, only k out of n acts need be performed successfully. For such a situation, the reliability of the sequence is:

$$P(\text{success}) = \sum_{i=k}^{n} \binom{n}{i} p^i (1 - p)^{n-i}. \quad (k \text{ out of } n)$$

The relationship between system reliability, P, and individual reliability, P, based on such a criterion is illustrated in Figure 1.4.

This view of behavioral acts and sequences leads into new insights into the origins of eusocial behavior through two remarkable theorems from the mathematical theory of reliability (Barlow and Proschan, 1975). First, it can be shown that the reliability of a system P is related to the reliability of its individual components, p, by a sigmoidal curve similar to that shown in Figure 1.4 for the k out of n function. That is, when the competence of individuals is low, the performance of the group is lower than that if the individuals acted independently. However, there is a threshold, p^*, of individual reliability above which group reliability exceeds that of individual reliability.[1] In this region, group selection will favor cooperative behavior.

An example will help to make this relation intuitively clear. Suppose that several persons were trying to stack a specified number of bottles one on top of the other in the middle of a room. If one person tried it from the floor up, and then another started fresh, and so on, the group result would equal the

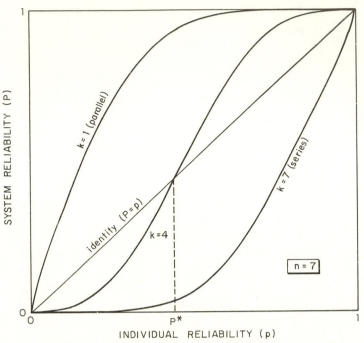

FIGURE 1.4. The reliability of a system (the probability that it performs an act correctly either once or on some designated number of times) depends on whether its units operate in series or in parallel. P is the probability that the entire system operates successfully (i.e., the system reliability) and p is the probability that an individual carries out its task successfully (i.e., the individual reliability). If the requirement is to perform the act correctly during at least k attempts our of n, the overall success (P) increases with individual competence (p) sigmoidally. For $n = 7$, the value $k = 7$ corresponds to a completely serial operation, while $k = 1$ corresponds to completely parallel operation.

average individual result. Now suppose that two or more persons work together. If their competence (p_i) is low, they will tend to cancel out one another's actions. The chance of making the right combination of correct movements (k/n) will be even lower than the chance of doing one thing correctly by itself, and fewer bottles will be stacked. However, as individual competence increases, the group will reach a point at which they are

14

able to put together the right combination of moves and balance a significantly higher number of bottles. This improvement will be greater still if they divide the labor. For example, one can hold and readjust the bottles already stacked, a second can put more bottles on top, while a third can fetch still more bottles to add to the attempt.

The potency of colony-level selection is also enhanced according to a second theorem from reliability theory, which states that replication at the subunit level is more efficient than redundancy at the system level (Barlow and Proschan, 1975). Put another way, if a designer is given two sets of "parts," it is better to build a single system with redundant components than to build two separate systems. The reliability of the redundant system—meaning the chance it will work at all—is greater than the reliability of the separate systems. This relation is illustrated in Figure 1.5.

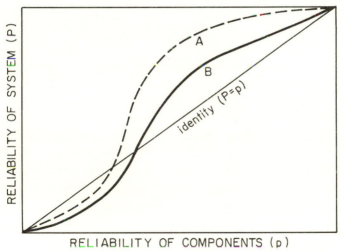

FIGURE 1.5. The reliability of a system is shown as a function of the reliability and redundancy of its components. *A*: System with redundancy at the subunit (components) level. *B*: System with redundancy at the systems level but not at the subunit level.

These results are not restricted to behavioral acts which either succed or not, as we have assumed above; graduated degrees of competence can be incorporated. More importantly, correlations between the behavior of individuals (i.e. the p_{ij}'s) can be included. These greatly increase the technical difficulties of computing the system reliability, but the theorems remain essentially unaltered.

1.4. THE EUSOCIALITY THRESHOLD

It is easy to see how the same principle can be applied to social insects. Groups of colony founding queens might easily attain levels of proficiency beyond the reach of a single individual. If modestly endowed with individual competence and the ability to divide labor, they could construct and defend the nest and rear offspring with greater effectiveness. In fact, such group superiority has been amply documented in some species of ants and wasps, and these data will be reviewed in Chapter Two.

The question then arises, why aren't all insects social? The answers appears to be that under a wide range of conditions the activity of the group yields fewer results than the *summed* efforts of an equivalent number of solitary individuals. To use the bottle example again, let us suppose that three competent players can stack six bottles, while one person can only stack three bottles. Even so, the three members of the team working singly can stack a total of nine bottles. If they were then each to be rewarded according to the total number of bottles stacked, their best strategy would be to work independently. In the same fashion, an individual's share of the rewards of an insect colony—that is, the share of the inclusive fitness—will under many circumstances be less than the profit to be obtained by working alone. In other words, the bottle-stacking team and the ant colony can produce a very superior single product, but the average reward to the individual team member may be smaller.

What, then, makes it possible to produce not only the superior single product but also a larger share of the profit than would be otherwise possible? It is a remarkable fact, which we will examine in greater detail in Chapter Three, that 12 of the 13 known groups of insects that independently evolved eusociality belong to the single order Hymenoptera (ants, bees, wasps); the thirteenth comprises the Isoptera (termites). A species of wooly aphid, *Colophina clematis*, has evolved a non-reproductive soldier caste but not the additional features of eusociality (Aoki, 1977). Part of the explanation seems to lie in the haplodiploid mode of sex determination, which characterizes the Hymenoptera and few other insect groups. Haplodiploidy results in females being more closely related to their sisters than to their brothers; in the case of a singly inseminated and outbred mother, the coefficient of relatedness of sister to sister is 3/4 and of sister to brother 1/4. The result should be a female-oriented society with strong altruism toward sisters, which is conspicuously the case in nature.

Several authors, including Lin and Michener (1972), Michener and Brothers (1974), Alexander (1974), and West-Eberhard (1975), have expressed dissatisfaction with this rather stark, algebraic explanation. They have urged the equivalent importance of various additional advantages of early eusociality, which are considered somehow to favor the Hymenoptera: the superiority of mutual nest construction and defense, the opportunities workers enjoy of ovipositing surreptitiously in the presence of the queen or even taking over when she leaves or dies, and the advantage the queen receives from simply controlling and exploiting the workers at the expense of their own genetic fitness. Michener, West-Eberhard, and their co-workers view these factors as being potentially auxiliary to kin selection, while Alexander favors the extreme hypothesis that exploitation is the primary force and kin selection negligible.

What is required next is an accounting system. We would like to be able to write something like a multiple regression equation incorporating each of the factors, in addition to kin selection, and yielding the inclusive fitnesses of the would-be queens and workers. When the averaged fitnesses reach a certain *eusociality threshold* (Wilson, 1976c), the species is likely to evolve all three of the basic traits of a "higher" social (eusocial) insect—cooperative brood care, division of the group into reproductive and sterile castes, and overlap of at least two generations.

There appear to have been two preconditions that carried the 13 phyletic lines across the eusociality threshold. The first is the enhancement of kin selection by haplodiploidy. Other kinds of arthropods, including many beetles, spiders, and orthopterans, build nests, manipulate objects skillfully, and care for their young, occasionally to an extreme degree; but none has attained the eusociality threshold. It is equally true that haplodiploid enhancement is not enough. A few solitary haplodiploid arthropods exist outside the Hymenoptera, including certain mites and thrips, the aleurodid whiteflies, the iceryine scale insects, and the beetle genus *Micromalthus*.

A second group that has crossed the eusociality threshold are the termites. The primitive termites, all of which are eusocial, and their closest living relatives, the presocial cryptocercid cockroaches, share a unique trait: they are the only wood-eating insects that depend on symbiotic intestinal flagellates. The flagellates are passed from old to young individuals by anal feeding, an arrangement that requires at least a low order of social behavior. It is reasonable to suggest that termite societies started as feeding communities bound by the necessity of exchanging flagellates and, in a sequence that is the reverse of hymenopteran social evolution, only later evolved social care of the brood (Wilson, 1971).

It is possible that other behavioral preadaptations to eusociality exist that can never be reconstructed. However, the reliability argument of the previous section suggests that the onset of group advantageousness is a rather sharp threshold. Beyond this point the effects of colony selection can rapidly

18

amplify the properties of eusociality since, with only one or a few reproductively active members, the colony itself can easily become the unit of selection.

1.5. CASTE AND ROLES

Thus, with the advent of eusociality, hymenopterous and termite species were in a sense "freed" to explore the advantages of concurrent activity, caste, and division of labor. Let us next examine the relation between caste and efficiency more closely. The members of all eusocial insect colonies are divided into reproductive individuals, the queens (and in the termites, the kings), and the sterile or semisterile workers. The workers in turn are often differentiated into groups of labor specialists referred to interchangeably as subcastes or as castes in their own right. According to the species, there may be one or more large major castes distinguishable from the more typically constructed minor workers, or an exceptionally small "minim" caste, or both. All termite species are differentiated into such physical subcastes. A minority of ant species display a similar physical polymorphism, in some cases to an extreme degree, but only a very few wasp and bee species are polymorphic and then only to a rudimentary degree. Regardless of the presence or absence of polymorphism, the workers of virtually all kinds of social insects pass through well defined life stages across which their labor roles systematically change. Hence there exist two modes of division of labor, one based on physical castes and the other on temporal castes.

A caste can be roughly defined as any set of individuals that performs specialized labor in the colony for sustained periods of time. Occasionally, it is characterized more narrowly as a set of individuals in a colony that is both morphologically distinct from other individuals and specialized in behavior. In this book we will employ the broader usage as the more convenient for the development of theory. Whereas a caste is a set of individuals, a role is a pattern of behavior that appears consistently in societies belonging to a given species. The behavior affects other members of the society, either directly through communication and physical contact or indirectly, by means of protection from predators or modification of the physical environment. Members of a society can fill more than one role. For example, an ant might forage for food at one moment and nurse a group of larvae the next. Ideally, the full description of all roles together, insofar as they can be meaningfully separated, will fully define the society. And the society can be said to be divided into castes when discrete subsets of its members, typically but not necessarily distinguished by physical traits or age, also consistently fill particular roles. Finally, to avoid a *reductio ad absurdum* it is necessary to exclude from this definition differences based exclusively on sex or differences that arise naturally from the asymmetry of the relationships between parents and their offspring.

To understand caste and division of labor would be to grasp the full inner machinery of the insect society and its evolutionary development as a response to particular environments. Up to the present time students of the subject have stressed the mechanisms of caste determination. The problem they have addressed is the proximate means by which the caste of individual members of colonies is determined. In particular they have asked: given the existence of males, queens, minor workers, and soldiers, what genetic and physiological events lead an immature insect to grow up into one of the categories as opposed to another? Thanks to diligent research conducted during the past twenty-five years, a sound knowledge of the caste determination mechanisms of about twenty species of ants, termites, and social bees has been accumulated. This information has been summarized by Wilson (1971), Schmidt et al. (1974), and Lüscher et al. (1977).

But surprisingly little work has been devoted to the actual behavior of the castes or the evolutionary forces that shaped them. Our understanding of the simple facts of division of labor is very limited compared to the knowledge of caste determination. Here is an entire world of natural history and ethology waiting to be explored. Even less well understood, and seldom even considered in systematic fashion, is the problem of caste proportions. To this day we do not understand why some ant species develop a soldier caste while others, often in the same genus, do not. Or why soldiers comprise 20 percent of the worker force in one species while in another, closely related, species they make up less than 5 percent. In addition to small (minor) and large (major) workers some species possess an intermediate-sized (media) caste; in others the intermediate form is absent. Some species have long-lived workers that display idiosyncratic behavior based at least in part on knowledge of the terrain; in others the workers are short-lived and more nearly homogeneous in their routines. Such distinctions can be multiplied at length from the literature.

The guiding proposition of our inquiry is that variations in caste structure and division of labor reflect differing adaptations on the part of individual species of social insects. Thus, caste is not just central to social organization; it should provide the key to the ecology of social insects, insofar as those insects differ from their solitary counterparts. We regard many of the principal processes of colony life, including communication, physiological caste determination, and trophallaxis, as subordinate to the evolution of caste. We postulate them to be the enabling devices by which labor is allocated and by which the colony as a whole precisely adjusts its relationship to the nest environs.

1.6. THE SUPERORGANISM AND FACTORY-FORTRESS

Two economic metaphors are useful for conceptualizing an insect society. Viewed from the surrounding environment the colony can be regarded as a superorganism, a large and diffuse ameboid entity whose ingestive apparatus, comprised of the foraging workers, moves back and forth in circadian pulses over the surrounding territory. Viewed from within, the colony is like a factory enclosed in a fortress. Its Darwinian fitness is measured by the number of reproductive forms it releases to initiate new colonies. In order to maximize its colonial reproduction, the colony must extract the greatest possible amount of energy from the environment and convert it efficiently into colony growth. At the same time the colony must protect its investment, and especially the welfare of the queen, from the enemies that threaten the nest at all times. To maintain the nest as a fortress is costly in time and energy. Defense cannot be perfect, some loss must be sustained. Ideally the level of defense is that which, combined with the foraging and nurturent strategies, yields a growth pattern enabling the colony to convert its capital of immature forms and adult storage tissue into the largest possible crop of reproductive forms.

The imagery of classical economics has been used here with a purpose in mind. Most of the major advances of evolutionary biology and the social sciences have come from drawing meaningful limits around and giving precise definition to sets of previously ill-defined notions, then classifying them, and finally setting up an accounting system by which they can be quantitatively analyzed. The biology of caste seems especially well suited for this mode of analysis. Earlier, one of us (Wilson, 1968) suggested the use of the term ergonomics to designate the quantitative study of work, performance, and efficiency in animal societies. Elementary linear programming models were invented to apply these quantities to the study of insect castes. The result was a series of general conslusions about the evolution of castes and caste ratios that seem reasonably well supported by empirical evidence. But the models were not operational, in the sense that they failed to provide procedures for experimental testing and suggested few direct links to the known ecology of the social insects.

Our purpose in the present monograph is to examine the problem in greater detail and to provide an array of definitions

22

and models that relate to measurable quantities. The keystone of our theoretical development will be a reliance on the concept of optimization in nature. We, like the geneticists who have considered the caste problem previously, recognize that the evolution of caste and division of labor is guided primarily by natural selection at the colony level. But this is by no means exclusively the case. Individual-level selection operates during queen competition at the time of colony founding (Hölldobler and Wilson, 1977a) and in the queen-worker conflict over the sex ratio and energetic investments in the new queens and males (Trivers and Hare, 1976; but see the critique by Alexander and Sherman, 1977). The interaction of individual-level and colony-level selection complicates the evolution of caste in ways that are just beginning to be explored.

Because optimization plays such a central role in our model building, we foresee an additional, more general benefit from the study of caste: a test of the concept of evolutionary optimization itself. If detailed and falsifiable hypotheses can be fashioned in advance of experimental research, they will serve to test not only caste theory but also the fundamental assumption of evolutionary optimization routinely made by biologists. If this assumption should fail in the social insects, which can be characterized as the "squid axon" of sociobiology, the debacle could have major repercussions for general evolutionary biology. A whiff of danger adds excitement to our subject and should help to attract more of the imaginative investigators that it clearly deserves.

SUMMARY

Colonies of social insects are distinguished from solitary insects less by novel behavior patterns than by increases in the scale and efficiency of their operations. The key to this improvement appears to be the ability to conduct activities concurrently instead of in series. When individual competence is low, it is advantageous to operate singly; but at a higher threshold level of individual competence, it becomes far better to operate concurrently in groups (Figures 1.3 and 1.4).

In spite of the potentially great rewards of group activity, advanced social life—eusociality—is known to have originated only thirteen times in evolution, and the vast majority of the hundreds of thousands of species of insects and other arthropods are entirely solitary. The evidence seems best explained by positing the existence of a eusociality threshold, defined as the level of inclusive fitness above which eusociality can evolve within species. Contributions to this fitness probably come variously from the superiority of mutual nest construction and defense, the possibilities of usurpation of the reproductive role within the colony, and the advantage the queen receives from controlling and exploiting workers at the expense of their own fitness. But at least one additional special contribution is needed. In the Hymenoptera this has evidently been the asymmetric degrees of relatedness resulting from haplodiploid sex determination, which give an advantage to rearing sisters instead of personal offspring. In the termites the special circumstance appears to have been the need to exhange symbiotic, cellulose-digesting flagellates.

Once species have passed the eusociality threshold, they are "freed" to utilize the full advantages of caste and division of labor. Castes are defined as sets of colony members that perform specialized roles over sustained periods of time; they are in addition often distinguished by membership in certain anatomical or age categories. Division of labor has been generally accompanied by a reduction in the overall behavioral repertories of individual caste members.

Because of the absence to this point of a true theory of caste, researchers have concentrated on the physiological factors that determine the caste of individuals rather than on the more fundamental role of the castes and meaning of caste ratios. Our entrée to caste theory is the systematic treatment of the colony as an economic system. We start by imagining the colony as a

24

fortress-factory, an entity that orchestrates its total activity to maximize the production of the reproductive castes capable of founding new colonies. From this viewpoint castes and division of labor are subject to optimization by natural selection at the colony level.

Our theory of caste thus pivots on optimization theory. There exists a possible reciprocal benefit: because of the relative rigidity of insect behavior, caste theory appears to be an especially favorable testing ground for the general concept of evolutionary optimization.

NOTE TO CHAPTER ONE

1. Compound systems are still more likely to contain thresholdlike transitions. For example, if the system is composed of n units, of which k must work, and each of these units is in turn made up of n components of which k must work, then this composite system will approach a step function as closely as desired.

CHAPTER TWO

Colony Life Cycles

Like many other complex phenomena, caste is profitably viewed as a hierarchy of processes, each level describable by its own models and each susceptible to dissection into the causal phenomena at the next lower level of organization. This is the conception by which the remainder of the book has been planned. Our account begins with a macroscopic view of the principal colony life cycles, including a brief consideration of the major selective forces that account for the observed variation. We next pass to a consideration of ergonomic efficiency at the various colony life stages, and then on downward to detailed definitions of labor roles and castes that can be utilized in ergonomic theory. Finally, we state the basic theoretical problem in terms of three socioecological functions that must somehow be related in order to provide a full evolutionary explanation of caste; these are the caste distribution function (CDF), the resource distribution function (RDF), and the defense distribution function (DDF). The success of the whole enterprise turns on whether the socioecological functions can be characterized in a way that permits the basic theory to be tested through laboratory and field studies.

2.1. THE STAGES OF COLONY GROWTH

For convenience we will define a "basic" colony life cycle as one that contains the following stages. A virgin queen departs from the nest in which she was reared, leaving behind her mother, who is the queen of the colony, and her sisters, who are either sterile workers or virgin reproductives like herself. In the case of termites, her father is probably still living as a royal consort to the mother queen, while some of her brothers serve as part of the sterile worker caste. The queen finds a mate during her nuptial flight and is inseminated. In the case of hymenopterans (ants, wasps, and bees) the two quickly separate and the male dies, but in the case of termites the partners stay together and cooperate in nest founding. The queen then finds a suitable nest site in the soil or plant material and constructs a first nest cell. Here she rears the first brood of workers. Soon after reaching the adult stage, these individuals take over the tasks of foraging, nest enlargement, and brood care, allowing the queen to confine herself to egg-laying. Over the coming weeks and months the population of workers grows, the average size of the workers increases, and new physical castes are sometimes added. After a period ranging according to species from a single warm season to five or more years, the colony begins to produce new queens and males. These reproductives set out to start new colonies, and the colony life cycle begins anew.

Substantial variation exists on this elementary theme, especially with reference to the mode of colony founding and the number of egg-laying queens that coexist during various stages of the life cycle. Figure 2.1 presents a classification of the variations and a minimal amount of the relevant terminology. *Monogyny* refers simply to the possession by a colony of a single queen, as opposed to *polygyny*, which is the possession of multiple queens. The founding of a colony by a single queen is referred to as *haplometrosis*; when multiple queens start a colony the condition is called *pleometrosis*. The term *metrosis* refers generally to this biological variable. Monogyny can be *primary*, meaning that the single queen is also the foundress; or it can be *secondary*, meaning that multiple queens start a colony pleometrotically but only one survives. In a symmetric fashion, polygyny can be primary (multiple queens persist from a pleometrotic association) or secondary (the colony is started by a single queen and

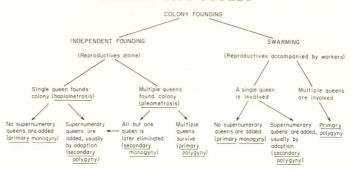

FIGURE 2.1. The "basic" colony life cycle is modified in various species of social insects by variations in the mode of colony foundation and the number of egg-laying queens that coexist in various stages of the life cycle. This diagram presents the several possibilities and the special terms employed by entomologists to describe them. (From Hölldobler and Wilson, 1977a.)

supernumerary queens are added later by adoption or fusion with other colonies). Finally, the mode of colony founding is subject to complicated variation among species. It can be accomplished by swarming (a process also called budding, hesmosis, or sociotomy), in which one or more reproductive forms depart with a force of supporting workers. Colony founding is frequently *claustral*, meaning that the queen seals herself off in a chamber and rears the first brood in isolation. This is the prevailing mode of independent colony formation in ants, although the queens of such primitive forms as *Amblyopone* and *Myrmecia*, in addition to those of a few more advanced species (in *Acromyrmex* and *Manica*), still forage outside their cells for food. Claustral founding is also generally practiced by termites but is absent in the bees and wasps.

Through all these variations runs a thread of common events. All life cycles can be easily divided into what appear to be three natural stages, each of which is shaped by a distinctive blend of individual-level selection and colony-level selection. The three periods can be conveniently designated the *founding stage*, the *ergonomic stage*, and the *reproductive stage* (see Figure 2.2).

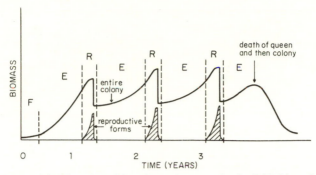

FIGURE 2.2. The colony life cycle can be conveniently divided into the three stages indicated here, each of which is molded by a different blend of individual- and colony-level selection and requires a separate analysis of caste and division of labor. *F*, founding stage; *E*, ergonomic stage; *R*, reproductive stage. With the release of the reproductive forms (virgin queens and males) in the breeding season, the colony is diminished in size and reenters the ergonomic stage.

At any point in the life cycle the colony growth rate must obey the following mass conservation balance:

$$\frac{dB}{dt} = P - M - D - F \qquad (2.1)$$

where B is the biomass of the entire colony measured, say, in calories; P is the colony biomass production rate; M is the total colony metabolic rate; D is the total mortality rate converted into calories; and F is the rate of release of reproductive forms destined to mate and to disperse to new sites, also expressed in calories.

During the founding stage, the independently founding queen rears a first brood of workers on her own. When the queen feeds the larvae on her own tissue reserves, the caloric content of the little group actually declines as the number of colony members is increased. Since the queen is the only adult present, there can be no division of labor unless the larvae metabolize foodstuffs in specialized ways and pass back nutrients to the queen (see Maschwitz, 1966; Ishay and Ikan, 1969).

29

When the first workers eclose as adults, the colony enters the ergonomic stage. Now there is a division of labor: the queen reverts mostly or exclusively to egg-laying and the workers assume the remainder of the tasks. The activities of the colony are exclusively ergonomic in the sense that they are concerned with work devoted to colony growth, rather than with colony-level reproduction or dispersal. After the colony reaches a certain "mature" size, to use the expression of Brian (1965a), it begins to produce virgin queens and males, initiating the reproductive stage. When the reproductive forms depart on nuptial flights, the mother colony either dies or is returned temporarily to the ergonomic stage.

Let us now examine the three colony life stages in more detail, with special reference to the levels of natural selection that affect them. A more detailed justification of this division of the cycle into discrete stages will be given later in this chapter.

2.2. THE FOUNDING STAGE

During colony foundation the crucial processes are dispersal to and colonization of acceptable nest sites. Caste is temporarily rendered less important in the daughter units. In the extreme case, independently founding queens must exist for a while as solitary insects. Their success depends on their ability to behave like the females of otherwise similar nonsocial species.

It is probably no accident that the queens of so many social insects pass through a solitary phase. Hölldobler and Wilson (1977a) have argued that independent colony foundation must be favored by natural selection under a wide range of conceivable ecological conditions. If entire colonies can be started by one or a few queens, mother colonies producing such females can deploy far more of them over greater distances than otherwise comparable colonies that reproduce by swarming. The reason is that each swarm necessarily drains off a substantial part of the original worker force, and its dispersal range is limited by the difficulties inherent in orientation and mobility of the swarm. Swarming is likely to be advantageous only if the survival rate of queens is overwhelmingly greater when they are accompanied by workers than when they proceed alone.

Furthermore, haplometrosis can be expected to be the preferred mode of independent colony foundation. Unless circumstances give a large advantage to founding in groups, each queen should attempt to start a colony well away from all possible rivals. Even when she joins a group, each queen should try to gain the position of exclusive egg-layer as soon as the first workers appear. The restitution of monogyny by dominance means that a queen choosing to become a member of a group of n founding queens has a $1/n$ chance of surviving to be the nest queen.

Finally, colony foundation can probably be expected to evolve so as to be claustral. Social insect workers suffer their highest mortality during foraging trips, and it is probable that the same is true of founding queens forced to leave their nests in search of food.

To summarize, a logical examination of the general properties of insect societies leads us to expect that social insect species have evolved in the direction of the most solitary mode of colony foundation, involving both haplometrosis and a claustral rearing of the first brood. Yet deviations from this expected pattern are frequent. Consider, for example, the differences between ants and social wasps. A fact in agreement with our line of reasoning is that most ant species are obligatorily haplometrotic and claustral. In contrast, among the wasps only the temperate-zone species of *Vespa* and *Vespula* are known to be entirely haplometrotic, and even the founding queens of these insects forage in order to obtain food for their growing larvae. *Belonogaster*, *Mischocyttarus*, and *Polistes* are sometimes haplo- and sometimes pleometrotic, while most or all of the Polybiini, containing 20 of the 26 known social wasp genera, reproduce by swarming (Evans and Eberhard, 1970; Spradbery, 1973).

31

Hölldobler and Wilson (1977a) have suggested the following simple explanation for the difference between the two groups of social insects, based on the mode of locomotion. When an ant colony moves to a new nest site, for example following a disturbance by a predator, it must walk to the site. The workers are wingless, and the queen must travel on the ground with them. Thus, queens can afford to engage in claustral colony foundation. They shed their wings following the nuptial flight, and in all but a few primitive species histolyze their alary muscles to nurture the first worker brood within a completely closed cell. Since they never need to fly again, the queens are able to take advantage of the surplus energy available in the alary muscles. Both wide initial dispersal and independent, claustral colony founding are within their grasp. Since these are advantageous techniques under most conceivable conditions, the majority of ant species have evolved to utilize them.

When wasp colonies are disrupted, however, they fly to a new nest site. Lengthy ground travel is not only unnecessary but would be disadvantageous for insects so fully adapted to life in the air. Because the queens must fly with them, these reproductive forms must "stay in shape" by not losing their flight muscles. The nest queens of *Vespa* lose their power of flight as they grow older, presumably because of the weight of their ovaries, but this is the exception rather than the rule in social wasps. Thus, wasp queens are less well equipped to be solitary foundresses. Since independently founding individuals find it disadvantageous to convert the alary muscles into energy for the brood, they must engage in the risky process of foraging for food. It appears to follow that wasp species should be more likely to rely on pleometrosis or even swarming, which is in fact the case.

There are other deviations from the haplometrosis rule. In a sizeable minority of ant genera and species, pleometrosis is at least an optional mode of colony foundation. In other words, some queens attempt to start colonies singly after the nuptial

32

flight while others of the same species simultaneously form groups for the same purpose. When this phenomenon was examined in the laboratory in *Lasius flavus* (Waloff, 1957), *Solenopsis invicta* (Wilson, 1966), and *Myrmecocystus mimicus* (B. Hölldobler and W. Bry, personal communication), it was found that groups have a higher survival rate and bring their first brood to maturity more quickly than do single queens. Because multiple groups of queens of all three species are found in nature, it is reasonable to conclude that pleometrosis confers a selective advantage. The phenomenon is entirely consistent with the theory, presented in Chapter One, that predicts a general superiority of concurrent operations by multiple units over sequential operations. However, not all species form in groups, so clearly there are countervailing costs which hold sway under certain circumstances.

After completing their nuptial flights, queens of *Lasius neoniger* and *Solenopsis molesta* alight on the ground in what appears to be a random distribution. Afterward they shed their wings and search for suitable nest sites during short excursions on foot (Wilson and Hunt, 1966). If this procedure is followed generally in pleometrotic ant species, and if queens associate with some probability greater than zero whenever they contact one another on the ground, then the frequencies of the numbers of queens in various groups should fit a truncated Poisson distribution (zero class omitted), with the "quadrat" width being roughly half the distance over which the queens attract each other. This appears to be the case in data recently gathered for *Myrmecocystus mimicus* in the field in Arizona by Hölldobler (Hölldobler and Wilson, 1977a).

The cofoundress ant queens appear to be entirely amicable toward each other, cooperating fully to rear the combined brood. Such is not the case for the pleometrotic associations of the wasps *Polistes* and *Mischocyttarus*. One individual comes to dominate the others and to take over as the exclusive reproductive. The subordinate foundresses sometimes continue to oviposit, but their eggs are promptly eaten by the dominant female (Eberhard, 1969; Jeanne, 1972). A similar dominance relation has been recorded in the primitively eusocial halictid bee *Lasioglossum zephyrum* (Brothers and Michener, 1974; Michener and Brothers, 1974). In both the wasps and the bees the subordinate females take over most or all of the nonreproductive tasks of the colony. In effect, they become the first members of the sterile worker caste. The colony has already functionally entered the ergonomic stage.

2.3. THE ERGONOMIC STAGE: QUEEN/WORKER RATIOS

When the first brood of workers reaches the adult stage (or the subordinate cofoundresses cease reproducing), the little colony undergoes a radical transformation. If the queen has been conducting the ordinary chores of the colony, she now stops to devote herself exclusively to egg-laying. The workers take over all the remaining tasks, including the feeding of the queen herself. For a few worker generations, the average number of which varies among species, no new reproductive forms are produced. Also, few if any reproductive individuals or alien workers are adopted from the outside. Thus, the colony has become a closed system devoted to its own exponential growth. In vespid wasps (Spradbery, 1965), honeybees (Bodenheimer, 1937; M. V. Brian, 1965a), and ants (M. V. Brian, 1953, 1957a; Wilson, 1974a), the rate of colony growth is determined primarily by the size of the active worker force. In bumblebees, it is determined by the number of eggs laid by the queen, which in turn depends on the number of cocoons and hence the number of workers destined to emerge in the near future (Anne D. Brian, 1951). In all of these cases colony size is likely to increase exponentially during the first few worker generations, with small to moderate irregularities occurring whenever the queen temporarily suspends egg-laying, or episodic food shortages

34

delay larval growth, or any age group suffers unusual mortality. Because most attempts at colony founding succeed only in sites that are at most sparsely occupied by other members of the same species, the new colony typically finds itself with a relatively rich foraging area in the immediate vicinity of the next. Later, as the density of foragers increases, the energetic yield per worker may fall to the extent that population growth slows down (M. V. Brian, 1965a).

Thus, the colony can be viewed in its middle stage as a growth machine: its hypothesized "purpose" is to proliferate workers as quickly and safely as possible. This function is implemented chiefly by division of labor—the right number of foragers to harvest the territory, the right number of nurses to stoke larval growth, and a sufficient but not excessive number of defenders to stand by for emergencies. The focus of the colony during this period is ergonomic; reproduction and dispersal are not its immediate concern, while new nest sites are sought only when the old ones become untenable or too small to hold the growing colony.

During the ergonomic stage competition within the colony is at a minimum (an analysis of the competition will be given in Chapter Three). However, the beginning of the stage, or more precisely the transition to this stage from the preceding, founding stage, is sometimes accompanied by hostile interactions. In the ants *Lasius flavus*, *Myrmecocystus mimicus*, and *Solenopsis invicta*, pleometrotic laboratory groups revert to monogyny when the first brood appears. The *Lasius* queens fight with one another and then break apart into single-queen units (Waloff, 1957). Those of *Myrmecocystus* form dominance hierarchies, with the supernumerary individuals eventually being driven out by the workers (B. Hölldobler, personal communication). When multiple *Solenopsis invicta* queens are introduced to queenless workers, the latter usually execute all but one (Wilson, 1966). The behavioral or physiological basis for dominance in *Myrmecocystus* and *Solenopsis* remains unknown. In the carpenter ants

35

Camponotus herculeanus and *C. ligniperda* large colonies often contain several queens, but these individuals are intolerant of one another and maintain territories within the diffuse nests, a condition referred to by Hölldobler (1962) as oligogyny.

It is an interesting fact, in apparent harmony with the natural selection argument offered by Hölldobler and Wilson, that no case is known in ants of pleometrosis leading smoothly to polygyny. In 22 replicates out of 23 followed in the laboratory by Hölldobler, one member of a pair of *Myrmecocystus mimicus* queens was quickly ejected by the first workers that eclosed. In the remaining case, both queens coexisted during two years of colony growth—then one was ejected. Similarly, only one group out of 20 queenless groups of *Solenopsis invicta* workers kept both queens of a pair offered them (Wilson, 1966). It is possible that the facultative oligogynous and polygynous conditions found in species of *Aphaenogaster*, *Camponotus*, *Formica*, *Prenolepis*, and some other territorial ants are derived from such pleometrosis. The appropriate studies of most pleometrotic species have not been undertaken, either in the field or the laboratory. But if the transition does occur, it is still likely that some of the supernumerary queens are ejected in later periods of colony growth, and that the frequency distribution of queen numbers is thereby modified in form. In short, there appears to be a "bottleneck" at the time of the first brood maturation that consists in many ant species of a reduction in the number of queens, most commonly to a monogynous state.

Pleometrotic colonies of the wasps *Mischocyttarus* and *Polistes* pass through a similar bottleneck, in the sense that the auxiliary queens who join the dominant foundress are reduced to the status of workers during the founding stage. Thus, we conclude that among social insects in general, colonies begun independently are also haplometrotic or, if pleometrotic, they either quickly reduce the arrangement to functional haplometrosis by interindividual dominance or else convert to secondary monogyny when the first workers appear.

36

Yet in spite of the bottleneck phenomenon, many species are polygynous in later stages of colony growth. Among the polybiine wasps, especially the species characterized by longer colony life, young queens are regularly adopted back into the colony of their origin after being inseminated, and frequently large populations of supernumerary laying queens are built up. Colony multiplication then occurs by swarming, in which one or more laying queens departs with a group of workers to start life in a new nest site (Evans and Eberhard, 1970; Spradbery, 1973). Although secondary polygyny (in addition to primary polygyny) in wasps thus seems to be a concomitant of colony multiplication by means of swarming, swarming in other social insects preserves monogyny or its close equivalent. In the honeybee *Apis mellifera* the mother queen departs with a portion of the worker force before the daughter queen emerges from her pupal cell. Young queens of the stingless bees (tribe Meliponini) reverse this procedure by leaving the mother queen behind with part of the worker force (Michener, 1974). In army ants one or more new queens are first reared to maturity in the company of the old queen, then the worker force splits in two, each moiety taking one queen. In the tropical American army ants of the genus *Eciton*, supernumerary queens produced this way are sealed off by small groups of workers and left to die (Schneirla, 1971).

Secondary polygyny also occurs commonly in parasitic ant species, where mating occurs in or near the nest of the host species and the host workers accept inseminated queens (Buschinger, 1974). It is also the rule in ant species in which colony boundaries are weak or nonexistent and local populations consist of networks of intercommunicating aggregations of workers, brood, and fertile queens. Examples from the latter category include the wood ant *Formica polyctena* of Europe, the mound-building ants of the *Formica exsecta* group, the Argentine ant *Iridomyrmex humilis*, Pharaoh's ant *Monomorium pharaonis*, and the "*microgyna*" form of *Myrmica ruginodis*. Such species have been referred to as unicolonial, or more precisely as forming unicolonial populations, in opposition to the more frequent multicolonial ant species in which intercolony recognition and aggression occur (Wilson, 1971). Although the build-up of queens usually results from the readoption of newly inseminated individuals following their nuptial flight from the same nest, the initial state of polygyny in the arboreal acacia ant *Pseudomyrmex venefica* comes from the fusion of small haplometrotic colonies (Janzen, 1973a).

Both the absolute number of queens and the queen/worker ratio in secondarily polygynous colonies vary greatly. In fact, Elmes (1973) has gone so far as to suggest that the supernumerary queens of *Myrmica rubra* exist essentially as parasites on the colonies that readopt them. They contribute fitness to the colonies, of course, by simply serving as the progenitrices of the new aggregates necessary for overall population growth. But with reference to the growth rate of local aggregates they represent an extra energetic load without being essential for the minimal oviposition rate required to keep all of the worker force busy. The frequency distribution of *M. rubra* aggregates containing various numbers of queens fits the lognormal, or more precisely the lognormal Poisson. Several hypotheses lead to this result. The simplest is that the newly fecundated queens return randomly to the ground but then are recruited by the colonies with an efficiency that is not constant but rather a random variable. The latter condition, for example, would obtain if the recruitment power of an aggregate were a linear function of its size.

The important point concerning the large variability in the queen/worker ratio in polygynous aggregates is the possibility it raises that the relation between the queens and workers has evolved in a nonergonomic fashion—that is, without reference to the fitness of the colony as a whole, much less to its efficiency as a growth machine. The question is then raised: why should the workers tolerate extra laying queens whose presence must be paid for in a smaller future crop of virgin reproductives? Hölldobler and Wilson (1977a) have noted certain ecological correlations that point to a contribution by the surplus queens to colony fitness above and beyond the obvious contribution to their own individual fitness. Fully polygynous ant species— those that are unicolonial or at least comprised mostly of colonies with multiple queens—fall into the following two sets characterized by unusual and very different adaptation syndromes.

(1) The first group of species is specialized on exceptionally short-lived nest sites. Such species, which include *Tapinoma melanocephalum*, *T. sessile*, *Paratrechina bourbonica*, *P. longicornis*, and certain *Cardiocondyla*, occupy local sites that are too small or unstable to support entire large colonies with life cycles and behavioral patterns dependent on monogyny. These sites include tufts of dead but temporarily moist grass, plant stems, cavities beneath detritus in open, rapidly changing habitats, and other places likely to remain habitable for only a few days or weeks. Colonies in this polygynous class are characterized by extreme vagility—a readiness to move when only slightly disturbed and the ability to swiftly discover new sites and to organize emigrations. Their colonies are also typically broken into subunits that occupy different nest sites and exchange individuals back and forth along odor trails. It is apparently the latter quality that gives polygyny a premium in opportunistic nesting. Because of the inevitably frequent fragmentation of the colonies, subunits probably lose contact with one another for long periods of time, and occasionally forever. Having enough reproductive females to service most or all of the subunits means that the colony as a whole can exploit the rapidly fluctuating environment in which it lives. The colonies of other kinds of ants are not fragmented in this manner, and consequently a single queen suffices as the progenitrix of each full colony.

(2) The second set of fully polygynous ant species is specialized on habitats—entire habitats, as opposed to mere

nest sites—that are long-lived, patchily distributed, and large enough to support large populations. Because such sites are relatively isolated, the propagules of the species adapted to them encounter a potential bonanza when they succeed in colonizing one. They stand a good chance of building up a substantial worker force before other propagules discover the same place. Hence they are able to proceed with the thorough, dense occupation made possible by polygyny and budding. The habitat is filled with a single "supercolony," or unicolonial population, sometimes containing millions of individuals, that may have descended from a single propagule.

Examples of the second category include the Allegheny mound-builder *Formica exsectoides*, which typically occurs in persistent grassy or heath-like clearings. Such habitats are relatively scarce and patchily distributed, and many are fully occupied by huge unicolonial populations of the mound-builders. The *microgyna* form of *Myrmica ruginodis* shows a similar preference for scattered, very stable open habitats in England, some of which are known to have persisted at the same localities for as long as 200 years. The *macrogyna* form, which is haplometrotic and monogynous, favors less stable but more widespread habitats (Brian and Brian, 1955). *Pseudomyrmex venefica* is specialized to occupy species of swollen-thorn *Acacia* that grow for long periods of time in areas of slow floristic succession. Because the acacias are able to expand into extensive thorn forests, the *P. venefica* colonies have the opportunity to build large unicolonial populations over a period of many years. Single populations may contain 20 million or more workers, rivaling the African driver ants (*Dorylus*) for the possession of the largest ant "colonies" in the world (Janzen, 1973a). Tramp species, those ants distributed widely by human commerce and living in close association with man, are typically polygynous. In a sense they can be said to have been preadapted for patchy but persistent and species-poor habitats created within man-made environments. Most are comprised of unicolonial populations and spread largely or entirely through budding-off of groups of workers accompanied on foot by inseminated queens. Examples include some of the more serious of the ant pests: *Monomorium pharaonis, Pheidole megacephala, Wasmannia auropunctata,* and *Iridomyrmex humilis.*

To summarize this consideration of the queen/worker ratio, monogyny appears to have been consistently favored in the evolution of the social insects. Although of clear advantage to the individual who takes over the role of progenitrix, monogyny is also ergonomically efficient with reference to the colony, at least to the extent that a single queen can supply a sufficient number of eggs to keep the remainder of the colony fully engaged in brood rearing. Even colonies of army ants and leaf-cutter ants, containing hundreds of thousands or millions of workers, are monogynous; the huge queens are simply geared to manufacture the large number of eggs required to keep the colonies growing. Polygyny creates a higher queen/worker ratio than absolutely necessary and is therefore energetically inefficient. It dilutes the investment of individual workers in the genetic fitness of the reproductive crop. It also makes the precise regulation of reproductive events within the nest more difficult. Moreover, because the pattern of readoption of newly inseminated queens appears to be random with reference to the adopting aggregates (see Elmes, 1973), the queen/worker ratio itself is highly variable. Thus, polygyny is not an ergonomically efficient phenomenon in the narrow sense; the necessary reduction in the absolute number of workers prevents the colony from reaching its theoretically possible maximum growth rate during the ergonomic phase. But as the evidence just reviewed appears to demonstrate, polygyny does contribute to both the colony survivorship and long-range colony growth of certain species by permitting their aggregates to survive and to reproduce at higher rates under special environmental conditions at which monogynous colonies are less than optimal. Hence polygyny appears to be favored by colony-level as

41

opposed to individual-level selection. This interpretation needs to be tested and extended by much closer studies of the life histories and ecology of polygynous species.

2.4. THE ERGONOMIC STAGE:
THE FIRST BROOD

Let us now examine the ergonomic stage of colony growth in species that start their colonies independently and are subsequently monogynous. To the extent that colony-level selection operates, worker subcastes and division of labor based upon them are expected to be directed entirely to the survival and growth of the colony, with no consideration being given to colony reproduction. This characterization will be formally examined later in the chapter.

In achieving maximum growth, the patterns of caste and division of labor can be expected to shift as the colony changes in size. For example, the members of the first brood of adult workers are frequently so small in size as to constitute a virtually distinct caste. Entomologists sometimes refer to these as nanitic, or dwarf forms. These are also characteristically timid in behavior but otherwise perform the same repertory of tasks as workers in older, larger colonies. In the case of dimorphic species, the first-brood nanitics possess the basic anatomical structure of the minor worker caste. Major workers usually do not appear until later, and even then typically average smaller in size. In the Australian termite *Mastotermes darwiniensis*, to take an extreme case, the first soldiers to appear are miniatures of those that guard larger colonies, and they also differ slightly in several bodily proportions (Watson, 1974).

We suggest that the small size and timidity of the first workers represent prudent features built into the investment strategy of the colony as a whole. A newly founded colony should strive to maximize the number of workers and their initial survival rate at the expense of everything else.

It is reasonable to hypothesize that there exists a minimum number of workers needed to accomplish an adequate performance in each of the vital tasks—a certain number to enlarge the nest, a certain number to nurse the second brood, a certain number to forage, and so forth. There should also exist an optimum number, above this minimum, since adult mortality is inevitable before the second brood is matured. The optimum number of nanitics can be defined as that above which the survival of the colony can no longer be significantly increased, and in fact is likely to be decreased. The argument is as follows. Because the biomass of adult workers that can be produced by the founding queen is very limited, it is efficient for the queen to divide it into many small workers. But this advantage is soon reversed, because to raise a great many such individuals would necessitate the production of excessively small nanitics unable to exploit food items and nest sites for which the species is adapted. As a result, there should be an optimum number, determined by the balance between the advantages of a larger initial worker force and the disadvantages of a smaller body size (see Figure 2.3). This hypothesis appears tractable to experimental testing with laboratory colonies.

The optimal nanitic size can be analyzed by the following general procedure. If s is a linear measurement of a nanitic worker (it can be total length or any convenient allometric measurement), the founding queen's allocation to the initial brood of N workers is constrained by a relation of the form:

$$E = \text{constant} \cdot N \cdot s^3 \sim (\text{cal/mass} \cdot \text{numbers} \cdot \text{mass}).$$

The choice of a few large workers as opposed to a larger number of smaller ones is a question of the trading of profitability against safety. For purposes of illustration let us consider the restricted case in which the decisive factors are two in number: energetic profit from foraging and energetic loss due to mortality from predators.

43

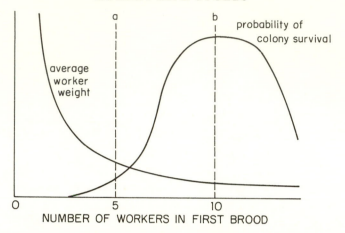

FIGURE 2.3. The first adult workers in a young colony are character-istically very small. The hypothesis represented in this diagram is that there exists a threshold number required to perform all of the vital tasks, and in addition a higher optimum number that compensates for average adult mortality prior to the maturing of the second brood. By producing very small workers at first, the incipient colony divides the available biomass of the first adult brood into the optimum number of nanitic workers. a = the number of workers in the first brood if the workers were as large as those in later broods. In this case, the number would be insufficient to bring a second brood to maturity. b = the optimum number.

A worker of size s can return an energetic profit to the new colony of $E(s)$ during the time it takes to rear the second brood. The larger the worker, the larger the size of the food particles it can carry. Its caloric profit will increase with size approxi-mately as s^3, since a particle commensurate with s has a caloric content proportional to its mass or volume. By this relation alone no substantial advantage is gained in foraging by manu-facturing either many small workers or a few large ones, since Ns^3 is approximately constant. However, if the size distribution of food items is not uniform, the size distribution of the worker force does matter. In nature (see Chapter Seven) the size dis-tribution of insect prey is in fact a unimodal curve which rises

44

from zero at a rate s^{α} ($1 \leqslant \alpha \leqslant 2$). Thus, within the class of workers best suited to handle prey of less than the modal size, larger individuals are likely to harvest more productively than smaller ones by an amount proportional to $s^{3\alpha}$.

The relationships between worker size and mortality due to predation are no doubt multitudinous and complex, and we will consider here only one of the simplest conceivable. As workers are made larger, and hence (in the special circumstances just cited) more productive, they are also made more vulnerable to predators due to enhanced conspicuousness. This enhancement might be on the order of s^2, that is, constitute a linear function of surface area. If predator attacks occur randomly, fitting a Poisson spatial distribution, then the probability of an individual surviving until the second brood matures is $e^{-\text{const.} \times s^2}$.

It follows that in this special case the expected energetic profit from N workers is the product of an increasing and a decreasing factor (see Figure 2.4). While few if any species are expected to be under such simple controls, the example can be used in a much more general way to demonstrate the likelihood of the existence of an optimal nanitic worker size. The important role of the balance between risk and return in caste evolution, which this case illustrates, will be considered at greater length in Chapter Six.

The small size of the incipient colony dictates that its members be relatively timid in behavior. Suppose that an encounter with a single enemy such as a group of foragers from an alien colony resulted in the loss of five workers. For a mature colony containing thousands of members, this cost might be tolerable, if it cleared enemy scouts from the territory on which the population has come to depend for food. But for an incipient colony of ten workers, the loss could be fatal. Furthermore, the potential gain from repelling territorial intruders is expected to be less, because the incipient colony is still living on a fraction of the available food supply yielded by the surrounding terrain.

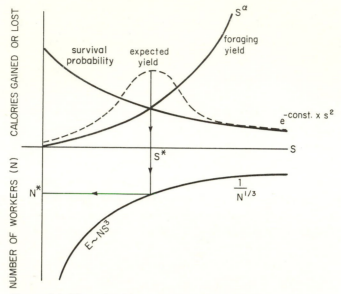

FIGURE 2.4. The optimal size (e.g., total length) of nanitic workers (S^*) and the optimal number of such workers in the first brood (N^*) are viewed as compromise solutions that maximize the average energy yield of the queen and first brood. The example shown here is based on a special case entailing simple relationships between worker size, foraging profit, and vulnerability to predators; a further explanation is given in the text.

It is in fact one of the most general rules of behavior in all social insects that the members of incipient colonies are more timid than the members of large colonies (see review in Wilson, 1971). Furthermore, within the large colonies, smaller workers appear to us to be typically more retiring than larger nestmates belonging to the same subcaste. However, this second generalization is certain only in the honeybee *Apis mellifera* (Kerr and Hebling, 1964), and it needs to be documented with reference to other species. To the extent that the behavioral differentiation exists, nanitics can be said to represent a "protocaste," even in species that are otherwise physically monomorphic. The timidity of the first brood of nanitics can be carried over

46

into a slight, statistical division of labor between small and large workers in more populous colonies later in the ergonomic stage. Smaller workers typically display a greater tendency to remain in the nest as nurses, while larger workers have a greater tendency to become foragers. Such differentiation represents a preadaptation to the emergence of a more distinct division into physical castes. An increase in size variation, especially when accompanied by stronger allometry, amplifies the division of labor in a minority of ant species. At first the division is probably in the direction of small nurses and large foragers. However, subsequent behavioral evolution has usually produced different patterns, with the largest individuals tending to become highly specialized to contribute to colony defense, seed milling, or food storage (see Chapters Six and Seven).

2.5. THE ERGONOMIC STAGE: LATER COLONY GROWTH

If the colony survives the precarious period during which the first and second worker broods are being reared, it is likely to enjoy a sustained phase of exponential growth. But this growth, like that in all populations, can be expected to slow with time and eventually to come to a halt. The few data on the actual course of colony growth suggest curves that are sigmoidal in form (Brian, 1965a). This is the expected result, but the underlying density-dependent controls are almost certainly more complex than those determining logistic growth in nonsocial species. The notion of "economies of scale," borrowed from economics, is useful in classifying the stages of colony growth (see also Figure 2.5).

(1) *Increasing Returns to Scale*

When the colony grows beyond the first brood of nanitic workers, it begins to reap new benefits based on larger population size alone. New physical castes are added—or at least

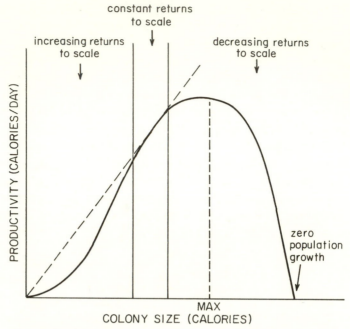

FIGURE 2.5. The colony productivity curves measure the net colony growth as a function of colony size. As the colony population increases, the total net yield of growth (measured in calories) can be expected to change in a way that conforms to a general pattern but varies among species. These imaginary curves hypothesize constraints in the ergonomics of most social insect colonies that result first in increasing returns, then in constant returns, and finally in decreasing returns to scale. The points of maximum colony growth and zero population growth fall within the period of decreasing returns to scale.

variance in worker size is increased. Consequently a wider range of food items can be harvested, soil particles of greater size can be moved to enlarge the nest, the colony can be defended by military specialists, and so forth. Efficiency in performing some of these tasks increases as the colony population grows, because masses of workers can now be recruited to handle larger food items, nest damage, or enemies. And since limited losses of workers in combat no longer threaten the existence of the entire colony, the foragers can be more aggressive in expanding their foraging range. At the same time the expanding colony is still small enough not to have outstripped its food supply. If the founding queen located a nest site far enough from larger colonies to have survived in the first place, the young colony is likely to find itself with a sizeable underexploited foraging area surrounding the nest site. All of these favorable circumstances should combine to confer increasing returns to scale. This means that for each unit of new energetic investment made by the colony—whether by foraging, defense, nest enlargement, or any array of other enterprises—the net growth of the colony increases more than proportionately. Doubling investment, for example, more than doubles return.

(2) *Constant Returns to Scale*

After a relatively short period of colony growth and increasing returns to scale the full ensemble of physical castes will be attained and the resource spectrum will be matched as closely as the colony is capable. So long as the food supply or some other density-dependent factor does not intervene, each unit of new energetic investment will now yield a proportionate amount of net growth in the colony as a whole. The colony will still grow, of course, but no longer at an accelerating rate. The incremental advantages of parallel operation are now being fully exploited (see Chapter One).

(3) *Decreasing Returns to Scale*

Ultimately all organized systems begin to decline in efficiency when they surpass a certain size. In the case of the growing colony, the carrying capacity of the territory must eventually be reached. If the defense mounted by neighboring colonies is not sufficient to halt the expansion of the territory, the energetic cost of its defense and of transporting food from its far reaches will become so great as to bring the colony to the point of zero net energetic yield. Also, the lines of communication can be strained in other ways. For example, if the colony becomes large enough to require multiple nest sites, travel between the sites is an additional costly and even dangerous activity. Indeed, workers of polydomous ant species are known to spend a significant amount of their time simply moving from one site to another, often along exposed trails. When the colony encounters a noticeable amount of resistance from such constraints, each unit of new energetic investment yields an ever smaller amount of net growth for the colony. Ultimately, net colony growth will fall to zero.

2.6. THE REPRODUCTIVE STAGE

If a monogynous colony were to increase its worker population to the point of zero population growth, it would be unable to reproduce, since total investment in workers means by definition that no production of virgin queens and males is possible. Consequently, at some point short of its maximum possible size, the colony should devote part of its production to the creation of virgin queens and males. The timing of the conversion is expected to vary among species according to the special adaptations the species have otherwise made to their environments. In the sections to follow we will examine this proposition in a series of detailed models.

What is the Optimal Reproductive Strategy of a Social Insect Colony?

The principal caste distinction in a social insect colony is between the reproductive forms (queens and males) and the nonreproductive workers. The fitness of a colony can be measured by the net rate at which it can produce reproductives during the colony lifetime. The key question then follows: what is the ergonomic strategy that produces the largest reproductive crop? To answer this question we will view the colony as a "machine" for converting resources into reproductives. The

50

workers forage and rear the queen's offspring, which are destined to become either reproductives or sterile workers. Without inquiring into the mechanism of caste determination one can ask: how should the colony resources be allocated between workers (labor) and reproductives (genetic capital) so as to maximize colony fitness? Colony-level selection will penalize the colonies that adopt suboptimal strategies.

Let us first address this question for the simplest case of social insects with an annual life cycle, such as bumblebees, polistine wasps, and vespine "hornets." In these forms only the new queens survive the winter, and the colony that produced her perishes along with the males who competed to fertilize her. In the spring the queen breaks her diapause, selects a nest site, and rears the first brood. Thereafter she remains inside the nest, laying eggs, while the workers forage to provision the colony and rear the subsequent broods. At some point during the season, reproductive forms are produced that are destined to leave the nest and mate, and then to found new colonies in the following spring.

In order to formulate the optimization problem quantitatively we consider a colony as consisting of only two castes. We assume that the colony's total effort—or productive capacity—is divided between the production of workers, W, and reproductives, Q. Since we are only concerned with the role of ergonomic strategies in colony evolution, all other contingencies are neglected that might affect colony growth (such as climatic extremes or parasites). Alternatively, if we consider an ensemble of colonies in the same habitat, then external selection factors act equivalently on each colony, and the ergonomic efficiency alone is the determining factor. The structure of the model is shown schematically in Figure 2.6.

At any time in the season the colony can devote a fraction $u(t)$ of its effort to the enlargement of the worker force and the remaining effort, $1 - u(t)$, to the production of reproductives. The colony fitness is measured by the net production rate of

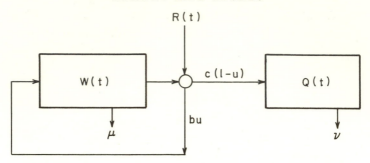

FIGURE 2.6. Schematic of the demographic model. The worker population $W(t)$ can reinvest a fraction $0 \leqslant u \leqslant 1$ of the resources gathered, $R(t)$, into producing new workers, and direct the remaining fraction $1\text{-}u$ into reproductives.

queens and males over the season. This is the same as the number of reproductives alive at the end of the season. Thus, the best reproductive strategy is to allocate $u(t)$ so as to maximize:

$$\mathcal{J} = \int_0^T \frac{dQ}{dt}\, dt = Q(T), \qquad (2.2)$$

where $\mathcal{J}$ is fitness and T is the length of a season.

The maximum fitness,

$$\mathcal{J}^* = \underset{0 \leqslant u \leqslant 1}{\text{Max}}\ \mathcal{J},$$

must be consistent with the colony's demographic trajectory; therefore, we must write a model for the population dynamics of colony growth. To do this we shall assume that, all other factors being equal, the production of new colony members is governed by the availability of resources (e.g., nectar and pollen) from the environment. This in turn depends on the proportion of workers engaged in foraging and the efficiency with which they are able to exploit those resources. We shall summarize both resource abundance and foraging capability in a "return function," $R(t)$, that gives the net rate at which resources are returned to the colony per forager. In order to determine the form of $R(t)$, a model for the foraging process must be constructed; we shall discuss this in Chapter Seven. For present purposes, however, we can treat $R(t)$ phenomenologically.

The equations governing the rate of worker production can be written in the form

$$\frac{dW(t)}{dt} = \text{Production} - \text{Mortality}$$

$$= P[W,uR] - M[W]. \qquad (2.3)$$

If we assume that worker mortality is constant and random then the second term is simply μW. The production rate, however, depends on the overall ergonomic efficiency of the colony in accumulating resources and converting them to live biomass. The simplest assumption is the one of linearity illustrated in Figure 2.7, which we expressed as

$$\text{Production} = bu(t)R(t)W(t). \qquad (2.4)$$

That is, production is proportional to both the fractional resource utilization uR and to the labor force W; b is the conversion constant. It follows that the demographic equation

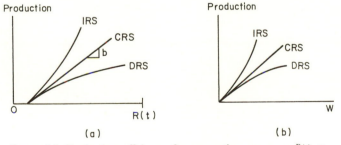

FIGURE 2.7. Production efficiency for converting resources $R(t)$ to worker biomass. (a) Production as a function of resource utilization; (b) production as a function of worker labor force. CRS = linear, constant returns to scale; IRS = increasing returns to scale; DRS = decreasing returns to scale.

governing the worker population is:

$$\frac{dW(t)}{dt} = bu(t)R(t)W(t) - \mu W(t) \tag{2.5}$$

$$W(0) = 1, \qquad W(t) \geqslant 0.$$

The initial condition $W(0) = 1$ implies that we are counting the founding queen as a worker since she forages to feed the first brood.

The production of reproductives also depends on the resource allocation $(1 - u(t))R(t)$ and on the worker force, $W(t)$; because of the size differences between workers and reproductives the conversion constant will be different.[1] Therefore, we can write for the reproductive population (see Figure 2.6):

$$\frac{dQ(t)}{dt} = bc(1 - u(t))R(t)W(t) - vQ(t), \tag{2.6}$$

$$Q(0) = 0, \qquad Q(t) \geqslant 0,$$

where v is the mortality rate of reproductives.

Thus, the optimization problem to solve is the time course of resource allocation $0 \leqslant u^*(t) \leqslant 1$ that maximizes $Q(T)$, subject to the demographic constraints given by Equations (2.5) and (2.6). In order actually to carry out the calculation it is necessary to determine the specific functional form of the return function, $R(t)$. This will be treated in detail in the foraging models of Chapter Seven. At this time, however, we shall anticipate the outcome by employing a generic distribution flexible enough to describe a wide variety of return functions. Such a function is the Beta distribution (Johnson and Kotz, 1970, p. 37).

$$R(t) \sim \beta(t;p,q) = \frac{1}{B(p,q)}\, t^{p-1}(T - t)^{q-1}. \tag{2.7}$$

By adjusting the parameters p and q the shape of $\beta(t)$ can accommodate each of the functional forms expected for $R(t)$, as shown in Figure 2.8.

54

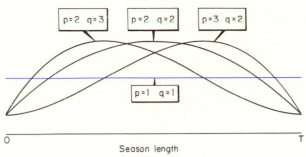

FIGURE 2.8. The shape of the return functions described by Equation (2.7) for several values of p and q.

In the results to follow the model parameters b, c, μ, ν, and T will be fixed by experimental data and the optimal reproductive strategy will then be computed as a function of the parameters p and q determining the shape of the return function. Although we shall have to adjust p and q to fit the observed data, the model's qualitative predictions will not depend on the precise form $R(t)$. Moreover, the return function required for quantitative prediction conforms to one's intuitive estimate, and indeed is subject to experimental verification. The model equations constitute an optimal control problem that can be attacked using conventional methods. The method of solution is briefly outlined in Appendix 2.1; a more complete discussion can be found in Macevicz and Oster (1976). The major feature of the solution, which is independent of the shape of the return function, is the qualitative nature of the optimal reproductive strategy. The maximum reproductive fitness is achieved by adopting a so-called bang-bang policy for colony growth.[2] *That is, in order to produce the maximum number of queens in a season of length T the colony should divide its reproductive effort into two distinct phases.* In the first phase the total reproductive effort is directed toward building up a large worker force; no reproductives are produced. Then, at a critical time, t_s, the entire reproductive effort is thrown into producing queens and males. Thus, the optimal allocation function $0 \leqslant u(t) \leqslant 1$ always has the appearance shown in Figure 2.9. $u = 1$ on $0 \leqslant t < t_s$ and $u = 0$ on $t_s \leqslant t \leqslant T$. In no case is it more efficient to produce workers and/or reproductives at an intermediate rate, $0 < u(t) < 1$. In Figures 2.10a and 2.10c we have plotted the data of Ishay et al. (1967) for the wasp *Vespa orientalis* and of Eberhard (1969) on the paper wasp *Polistes fuscatus*. The "explosive growth" strategy is the most striking feature of the demographic trajectory, and is typical of many annual eusocial insects (e.g., Eberhard, 1969; Ishay et al., 1967; Plowright and Jay, 1968; Röseler, 1974; Cumber, 1949; Richards and Richards, 1951).

The data on *Polistes* (Figure 2.10c) could not be fitted with an exponential growth model; a logistic density dependent term is consequently added to the worker production term so that the equation for worker production becomes:

$$\frac{dW}{dt} = b_0 u R W (1 - b_1 W) - \mu W. \tag{2.8}$$

The model equations are shown in Figure 2.10d.

In order to use the model to compute the actual switching time, t_s, we must choose a specific form for the return function. In the simplest case of exponential worker growth and constant

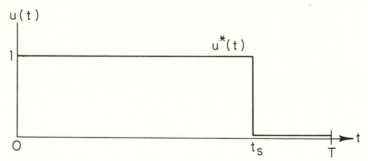

FIGURE 2.9. The form of the optimal solution $u^*(t)$ to the control problem described by Equations (2.2) through (2.7) using system parameters obtained from data in Ishay et al. (1967) and Eberhard (1969).

return function t_s can be computed explicitly (cf. Appendix 2.1):

$$t_s = T - ln\{1/[1 - (\mu - v)/bR]\}^{1/(\mu - v)}. \qquad (2.9)$$

However, by varying the parameters p and q in the β-distribution model for the return function we can compute numerically the switching time for the logistic model and for any realistic $R(t)$. Such a plot is shown in Figure 2.11 for the data corresponding to Figure 2.10a; Equation (2.9) is the value at $p = q = 1$. In general, we see that later switching times are promoted by (1) lower worker mortality rates and/or higher offspring mortality rates, and (2) resource spectra that peak toward the end of the season.

Figures 2.10b and 2.10d show the demographic trajectories predicted by the basic model. These curves were generated by first obtaining the logistic parameters $\{b_0, b_1, \mu, v\}$ from the data, then solving the optimal control problem for the switching time as a function of the resource parameters p and q; finally, the shape of the return function was adjusted to the form shown by the dashed curve.

A qualitative appreciation of the nature of optimal strategy can be gained by examining the special case of Equation 2.9 when $v \to 0$, i.e., negligible offspring mortality before the season's end. Figure 2.12 shows a plot of t_s as a function of μ for several season lengths. Notice that as $\mu \to 0$, $t_s \to T - 1/bR$. That is, the switch to queen production occurs approximately one worker lifetime before the end of the season. This illustrates a quite general principle which turns up in a number of different biological settings: for example, the differentiation of immuno-competent lymphocytes into plasma cells (Perelson et al., 1976) and the switch from vegetative to reproductive growth in annual plants (Denholm, 1975; Mirmirani and Oster, 1978). In abstract terms the optimization problem can be stated thus: given the choice between "reinvestment" in the means of production (e.g., workers) or in using the means of production to manufacture a terminal product, what is the optimal allocation

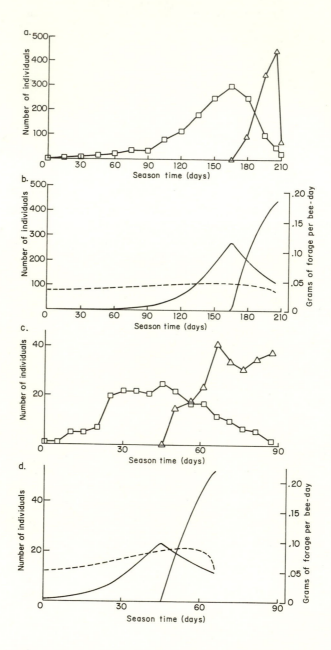

FIGURE 2.10. The observed and theoretical population curves for *Vespa orientalis* (a,b) and *Polistes fuscatus* (c,d) [a and c redrawn from Ishay et al. (1967) and Eberhard (1969).] □ : worker population levels. △ : reproductive population levels. *Dashed curves*: return functions used for computing the solution to the optimal control problem. The parameters used for computing each of the theoretical curves are as follows: *Vespa orientalis*: $b_1 = 0.0013$, $r_0 = 0.0535$, $c = 1$, $p = 28$, $q = 1.6$, $\mu = 0.022$, $T = 205.0$, $K = 3/2$ m, $b_0 = 0.5$, $t = 0$; *Polistes fuscatus*: $b_1 = 0.25$, $r_0 = 0.1$, $p = 2.8$, $q = 1.4$, $\mu = 0.034$, $c = 1$, $T = 66.0$, $k = 3/2$ m, $b_0 = 0.5$, $t = 0$. (From Macevicz and Oster, 1976.)

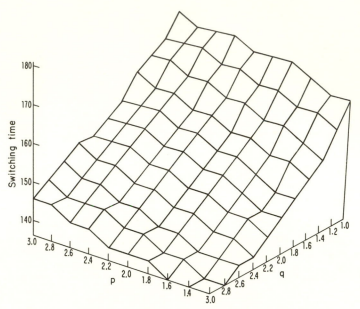

FIGURE 2.11. The switching times of the theoretical population of *Vespa orientalis* for various forms of the return function. The shape of the return function is determined by p and q. Large p and small q represent a return function which peaks toward the end of the season, and large q and small p represent a return function which peaks towards the beginning of the season. (The apparent irregularity of the surface is due to the coarseness of the grid used for numerical calculation.) (From Macevicz and Oster, 1976.)

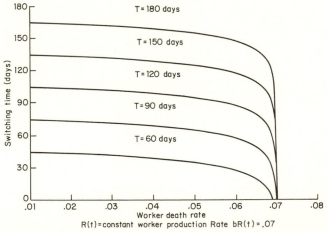

R(t)=constant worker production Rate bR(t)=.07

FIGURE 2.12. Sensitivity of the switching time from the ergonomic to reproductive stages in annual colonies as a function of worker death rate. (From Macevicz and Oster, 1976.)

of effort so as to produce the most product in a fixed time period? One might suppose that some intermediate policy of partial reinvestment would be most productive. However, we have seen that the optimal policy is to reinvest until the last possible moment (i.e., one mean life-time before the end of the allocated time) then switch entirely to product manufacture. No "splitting of the effort" yields higher returns. One can show that the switching time is sensitive to parameter changes only over a restricted region. Within most of the parameter range the switching time is only weakly and linearly sensitive to changes in birth and death rates.

The dependence of the switching time on the shape of the return function suggests that we examine the allocation of resources between workers and reproductives. The fraction of the total season's resources allocated to reproductives can be computed from:

$$f(p,q) = \int_{t_s}^{T} W(t)R(t;p,q)\ dt \Big/ \int_{0}^{T} W(t)R(T;p,q)\ dt. \quad (2.10)$$

This function is plotted in Figure 2.13 for the data of Ishay et al. (1967) as fitted by the logistic model. Together with Figure 2.11 it shows the relationship between the temporal pattern of the resource spectrum and the fraction of resources allocated to reproductives. We see that if the return function peaks toward the end of the season then the reproductives receive a greater proportion of the season's profits. This implies a greater ergonomic efficiency with respect to reproduction.

It is remarkable that so large a percentage of the resources gathered over a season are devoted to producing reproductives. This represents an ergonomic efficiency comparable to many man-made machines.

The major result of the above discussion is that for colonies in the region of constant returns to scale in their growth, the post-founding colony life cycle can be divided into two phases. The period $0 \leqslant t \leqslant t_s$, during which the colony reinvests its resources into an enlargement of the worker force, is appropriately called the ergonomic phase, as we suggested earlier.

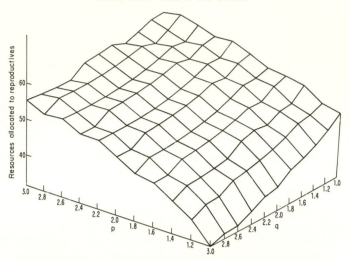

FIGURE 2.13. The percentage of resources allocated to the reproductives in the course of the season as a function of the shape of the return function [i.e., p and q in Equation (2.7)]. The area under the return function curves was held constant for each choice of p and q. Thus, the shape of the return function (independent of its magnitude) affects the manner in which the total resources accumulated by the colony are allocated between worker production and reproductive production.

During this interval selection is expected to act strongly to increase colony ergonomic efficiency. As the remainder of the book will make clear, caste differentiation is a powerful technique for enhancing ergonomic efficiency. Thus, during the ergonomic phase the selective forces that promote reproductive efficiency act to promote caste differentiation as well. The success of the reproductive phase is the payoff for colony efficiency during the ergonomic phase. During the reproductive phase selection acts differently on the queen and on the egg-laying workers, leading to a conflict of interest over the colony's reproductive policy. This interesting complication will be considered in detail in Chapter Three.

One assumption implicit in the above model deserves comment. With reference to strategic modeling no assumptions were necessary concerning the method by which a colony "computes" the optimal switching time, t_s. That is, the demographic model formulated here does not explicitly address the physiological mechanisms that actually control the switch in production and its timing, such as circadian clocks or worker/larva ratios.[3] Our modeling philosophy implicitly assumes that whatever the optimal strategy the population possesses sufficient genetic flexibility to implement it somehow and in fact has done so under pressure from natural selection. This is a dangerous assumption in general since there are several reasons why an optimal strategy may not be achieved. We shall elaborate on this potential defect in Chapter Eight.

Finally, we have assumed that the conversion efficiency was proportional to both the labor force and the resource utilization (i.e., constant returns to scale in Figure 2.7). It can be shown that increasing returns to scale also lead to optimal strategies which are bang-bang in nature. However, decreasing returns to scale or density dependence can produce graded strategies, consisting of simultaneous production of workers and reproductives. This suggests that large or perennial colonies, whose optimal strategy will be discussed in a moment, may find it selectively advantageous to adopt a mixed strategy. This point merits further theoretical and empirical study.

Perennial Colonies: Discounting for the Future

The optimal reproductive policy of colonies that last for more than one reproductive episode cannot be the same as the optimal policy of single-episode species. Clearly, maximizing the number of reproductives produced in season i, that is

$$\sum_{i=1}^{L} Q(T_i),$$

(where L = lifetime of the queen) neglects the possibility that the colony itself may not survive to the next breeding season. Thus, reproductives projected for manufacture in season $i + 1$

63

are not as valuable as ones produced in season i and must be discounted by the colony survival probability. The multiseason fitness criterion is consequently

$$J^* = \text{Max} \sum_{i=1}^{L} Q(T_i) \prod_{j=1}^{i} p_j. \tag{2.11}$$

If all of the annual colony survival probabilities, p_j, are equal,

$$J^* = \sum_{i=1}^{L} Q(T_i) p^{i-1}$$
$$= Q(T_1) + p Q(T_2) + p^2 Q(T_2) + \cdots \tag{2.12}$$

This fitness criterion can be interpreted as maximizing the *expected* number of reproductives produced over the colony lifetime (analogous to the net reproductive rate $\sum l_x m_x$ employed by demographers). For example, $p Q(T_2)$ is the expected number of reproductives produced in the second season (i.e., $p Q(T_2) + (1 - p) \cdot 0$, since no queens are produced if the colony fails to survive). Thus, p is just the system reliability discussed in Chapter One.

Throughout our considerations we have assumed that the population of colonies is constant. During the colonization of a new habitat, when the total number of colonies is increasing, an additional discount factor must be included in J to account for the added value of reproductives produced during the early stages of colonization.[4] Since our optimization models are designed to infer strategies evolved over many millenia we are probably justified in neglecting the colonization of new habitats in this context. (However, in Chapter Three we will study a situation in which colonization effects can be critically important.)

Thus, the simplest multiseason optimization model is

$$\text{Max}_{0 \leq u \leq 1} \sum_{i=1}^{L} Q(T_i) p^{i-1}, \tag{2.13}$$

subject to the demographic constraints

$$\frac{dW(t)}{dt} = bu(t)R(t)W(t) - \mu W(t); \qquad W(0) = 1 \qquad (2.14)$$

$$\frac{dQ(t)}{dt} = c(1 - u(t))R(t)W(t) - \nu Q(t); \qquad Q(0) = 0 \qquad (2.15)$$

$$W(t), Q(t) \geqslant 0; \qquad 0 \leqslant u \leqslant 1. \qquad (2.16)$$

The solution to this problem is illustrated graphically in Figure 2.14 (see also Appendix 2.1). Each season is partitioned into a phase of ergonomic growth followed by a phase of reproductive output, exactly as in the single season case. That is, the bang-bang strategy is optimal within each season.

The additional parameter in the perennial model is the probability of colony survival between reproductive episodes, including for example overwinter survival. Fixing all other parameters, p determines the number of seasons with a switch as well as the switching time within each season, t_{si}. This is shown in Figure 2.15. Two points are worth noting.

(1) Colonies which on the average last many seasons (that is, have a large p) will tend to commence producing reproductives later in their life cycle (Figure 2.15a). High growth rates also

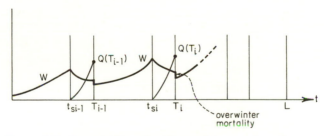

FIGURE 2.14. Perennial life cycle with annual decreases in worker population due to overwintering mortality and partial conversion of resources to the manufacture of reproductive forms.

65

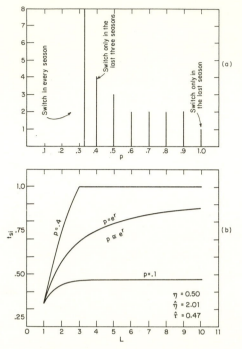

FIGURE 2.15. (a) The number of seasons with a reproductive switch as a function of the colony reliability, p. As the probability, p, of surviving over winter decreases, colony reproduction commences earlier in the life cycle. In very uncertain environments, where p is quite low, reproductives are produced in the very first season. Conversely, if the average colony overwinter survivorship is high, then reproductive production should be delayed, allowing the colony to build up a larger production facility. (b) Optimal switching times as a function of queen lifetime, L. The switching times increase as the colony lifetime increases but quickly approach a limiting value dependent on the colony survivorship, p. (From Mirmirani and Oster, 1978.)

promote delayed reproduction. Intuitively this seems to make good sense: small p means a short average colony lifespan, with the result that the delay of reproduction is dangerous. Conversely, a long expected lifetime means that a higher total reproductive output can be achieved by first accumulating a large worker force. This option is analogous to the single-season case.

(2) The switching times t_{si} increase as the maximum lifetime of the queen increases (see Figure 2.15b), but they approach an asymptotic value in the case of very long-lived colonies.

Available empirical data appears too scanty to test these predictions.

In this model we have treated p as a constant that measures the colony's capacity to cope with environmental uncertainty. Thus, p, the system reliability, depends on the caste ratios of the colony, reflecting in particular the efficacy of the soldier castes. In Chapter Six, when we construct more detailed models of caste structure this relation will be made more explicit. At this point the important step has been to establish the connection between caste structure and overall colony reproductive fitness.

The two principal results of the foregoing discussion are as follows:

(1) The colony life cycle is usefully divided into three more or less distinct stages: the founding phase, the ergonomic phase, and the reproductive phase. This partition is needed because the inferred forces of natural selection act differently during each phase. We will discuss the founding phase briefly in the next section and the reproductive phase in Chapter Three. Because the forces promoting caste evolution act most strongly during the ergonomic phase, the bulk of the analysis will be devoted to elucidating the ergonomics of caste evolution.

(2) The connection between the internal caste structure of the colony and the reproductive success of the colony is contained in the reliability measure p. To optimize the caste ratios is to optimize the colony's reproductive output. Subsequent arguments will demonstrate that the evolutionary process involves a trade-off between foraging productivity, which is contained in the return function $R(t)$, and the system reliability p.

How Robust is the Bang-Bang Strategy?

In both the single-season and the many-season models the optimality of the bang-bang strategy depends on several conditions which may not be generally valid:

(A) *Convexity of the conversion efficiency.* Constant or increasing returns to scale (convexity) promote discontinuous strategies while decreasing returns to scale (concavity) promote graded strategies; in the latter case the optimum procedure is to produce workers and reproductives simultaneously, as shown in Figure 2.16. The breadth of the sigmoidal region depends on the degree of convexity in the conversion efficiency curve.

(B) *Stochastic effects.* The basic model developed here is deterministic. It can be demonstrated that stochastic variation in the system parameters will always promote graded control; several examples of this phenomenon will be presented when caste differentiation is explored in Chapters Five and Six. The magnitude and importance of nonlinearities and stochastic influences on reproductive strategies is an empirical question worthy of additional study.

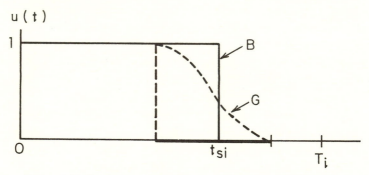

FIGURE 2.16. Reproductive strategy in season i. $u(t)$ = fraction of resources devoted to production of workers. Curve B: "bang-bang" or "all-or-none" strategy characteristic of constant or increasing (convex) returns to scale. Curve G: graded control characteristic of decreasing (concave) returns to scale (i.e., nonlinearity in u), stochastic variation in the parameters, and/or density dependence in the demographic equations.

(C) *Nonlinear state questions*. The demographic equations that constrain optimization were assumed to be linear.

2.7. GROWTH VERSUS CONSOLIDATION

From a consideration of system reliability it can be inferred that colony strategies will vary with colony age and size. Imagine, for example, that a colony of size *A* as depicted in Figure 2.17a has the option of investing its resources into the production of more workers and hence the expansion of the colony size for *B*. Alternatively, it can forgo expansion and

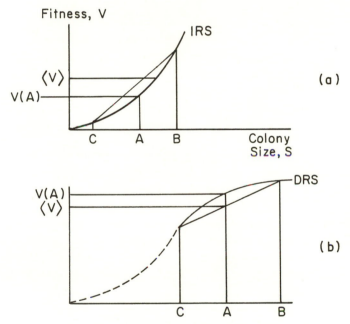

FIGURE 2.17. At increasing returns to scale (IRS), colonies will generally find it more profitable to invest in more workers and thus continue accelerated colony growth, upper diagram (a). At decreasing returns to scale (DRS), colonies will increase their mean fitness to a larger degree by investing in soldiers and hence consolidating their previous gains. See further explanation in text.

consolidate its gains by manufacturing a larger soldier force. If it chooses expansion, then it is likely to achieve its goal of B with some probability p, whereas it faces devastation to C with probability $1 - p$. The expected gain in fitness, V, is $\langle V \rangle = pS_B + (1 - p)S_C$. Over many trials gambling on expansion proves superior to consolidation since $\langle V \rangle > V(S_A)$. That is, colonies in the IRS (Increasing Returns to Scale) stage of colony growth will be selected for *Lebensraum* strategies. Conversely, colonies in the DRS (Decreasing Returns to Scale) stages will be selected to be more conservative since, as shown in Figure 2.17b, the expected gain in fitness $\langle V \rangle$ is less than $V(A)$. In general, convex fitness curves (i.e., $V'' > 0$) are associated with risk taking while concave fitnesses will select for risk aversion. The notion of risk aversion will play a key role in later models of caste composition. Density dependence can produce graded control, even though the logistic term required to fit the data of Ishay et al. (1967) on Oriental hornets did not do so. For example, the mortality of reproductives will generally depend on the size of the worker force, since one of the functions of the worker force is to protect the nest from predators. We can model this effect by replacing v in Equation (2.15) by a decreasing function of W, for example:

$$\frac{dQ}{dt} = c(1 - u)RW - ve^{-kW}Q. \tag{2.17}$$

For this circumstance it can be shown that the optimal strategy is graded with the length of the overlap region depending on the intensity of predation and the effectiveness of the workers' protective capabilities, both of which are incorporated in the single parameter k.

SUMMARY

The overall plan of the book is first presented. Following a general account of the life cycle of insect colonies, an exposition will concentrate first on the processes involved in colony ergo-

70

nomics, then on the division of labor upon which the ergonomics is based, and finally on the relation between the division of labor and the particular environmental circumstances in which species find themselves.

The typical colony life cycle appears to divide naturally into three periods, which we have labeled the founding, ergonomic, and reproductive stages. In each stage the colony is subject to a different blend of colony-level and individual-level selection. The reproductive stages are concerned to a large degree with mating and dispersal; during the ergonomic phase the colony becomes a highly coordinated "growth machine" seemingly designed to maximize the size of the worker population prior to the production of the virgin queens and males. Thus, the ergonomic stage, which occupies most of the life of the colony, is the period in which colony-level selection is paramount and division of labor can be expected to be the most complex and closely programmed. Possible changes in the rates of colony production as a function of colony size are examined, and their effects on the timing of the manufacture of reproductive forms roughly predicted.

A fundamental distinction is made between the numbers of queens founding colonies (metrosis) and the number occurring in colonies generally (gyny). The difference, which affects many aspects of the organization and ecology of the colony, appears to be based on the shift from entirely individual-level selection to primarily colony-level selection as the colony passes from the founding stage to the ergonomic stage of its development.

APPENDIX 2.1

The subject of dynamic optimization is quite technical, and it would be far beyond the scope of this book to attempt an exposition of the theory. Intriligator (1971) presents a clear and elementary treatment of the optimization tools we shall employ in this book. The mathematical steps leading to the conclusions cited in the text are found in Macevicz and Oster (1976) and Mirmirani and Oster (1978). The class of optimization models which the theory addresses is the following. By manipulating a set of "control" parameters $\mathbf{u} = (u_1, u_2, \ldots, u_m)$ (e.g., fractional resource allocations) we seek to maximize *over time* some fitness criterion, $\mathcal{J}$, which can generally be written:

$$\mathcal{J} = \int_0^T \Psi(\mathbf{x}, \mathbf{u}, t) \, dt + F(\mathbf{x}(T), u(T)),$$

where $\mathbf{x}^{(t)} = (x_1^{(t)}, x_2^{(t)}, \ldots, x_n^{(t)})$ are variables describing the system state (such as the population level). The values of the state variables $\mathbf{x}(t)$ are constrained to obey a set of differential equations of the form:

$$\frac{d\mathbf{x}(t)}{dt} = \mathbf{f}(\mathbf{x}, \mathbf{u}, t), \qquad \mathbf{x}(0) = \mathbf{x}.$$

Both the state and control variables are also constrained by certain inequalities, for example:

$$\mathbf{x}(t) \geqslant 0, \qquad 0 \leqslant \mathbf{u}(t) \leqslant 1.$$

The method of solving for the optimum reproductive strategy is a generalization of the technique of Lagrange multipliers. One first defines a function $H(x, u)$ to be maximized by:

$$H \equiv \Psi - \lambda^T \mathbf{f},$$

where $\lambda^T(t)$ are functions that obey their own set of dynamical equations defined by

$$\lambda = -\nabla_{\mathbf{x}} H,$$

where $\nabla_{\mathbf{x}}(\cdot)$ is the gradient with respect to the $\mathbf{x}$ variables only. The boundary conditions on the λ_i's are determined by requiring the state trajectory to arrive normal to the "terminal manifold" defined by

$$F(\cdot) : \lambda_i(T) = \frac{\partial F}{\partial x_i(T)}.$$

For the reproductive strategy model the Hamiltonian is

$$H = (\lambda_2 - \lambda_0)cR(t)W(1 - u) + \lambda_1[b_0R(t)W(1 - b_1W) - \mu W]$$

$$= b_0R(t)W\left[\lambda_1(1 - b_1W) - \frac{c}{b_0}\right]u + \text{terms not involving } u,$$

and the adjoint equations are

$$\dot{\lambda}_0 = 0$$
$$\dot{\lambda}_1 = -(\lambda_2 - \lambda_0)cR(t)(1 - u) - \lambda_1[b_0R(t)(1 - 2b_1W)u - \mu]$$
$$\dot{\lambda}_2 = 0.$$

Since H is linear in the control, u, in order for H to be maximized, u must take either its maximum or minimum values, ($u = 0$ or 1) depending on whether the quantity $\sigma(t) = [\lambda_1(1 - b_1W) - (c/b_0)]$ is positive or negative. During the evolution of the system, σ will start out positive, and become negative at some time t^*, at which point the control will switch discontinuously from $u = 1$ to $u = 0$. The actual calculation of t^* involves solving the state and adjoint equations, so that one can substitute explicit expressions for W and λ_1 into $\sigma(t)$. Then, by setting $\sigma(t) = 0$, one can solve for the optimal switching time

$$t^* = T - \frac{1}{v - \mu} \ln \frac{bR}{v - \mu + bR} \cdot$$

(See Macevicz and Oster [1976]; and Perelsen, Mirmirani, and Oster [1976, 1978].)

NOTES TO CHAPTER TWO

1. If the conversion factor b is measured in workers per gram (or calorie) of forage, then c depends only on the worker-to-reproductive weight ratio.

2. The term "bang-bang" is used in control theory to denote a control strategy that is all off or all on.

3. Synchronous nuptial flights clearly enhance mating success, so that any colony whose nuptial flight deviates significantly from the population nuptial flight will be severely penalized. However, the

model demonstrates that the bang-bang policy is ergonomically optimal as well.

4. The situation is analogous to discounting the future value of money by the interest rate. The present value of a sequence of payments $P_0, P_1, \ldots, P_L$ due in years $0, 1, \ldots, L$ is

$$\sum_{i=1}^{L} P_k (1 + \alpha)^{-i},$$

where α is the interest rate.

CHAPTER THREE

The Genetic Preconditions for Caste Evolution

One of the most suggestive facts concerning eusociality in the insects is its restricted phylogenetic occurrence. The condition has originated a minimum of twelve times in the order Hymenoptera and only once elsewhere (the termites, comprising the order Isoptera), despite the fact that the Hymenoptera comprise only about 15 percent of living insect species (Wilson, 1971). In 1964 W. D. Hamilton opened a new chapter in sociobiological theory by noting that the haplodiploid mode of sex determination in the Hymenoptera causes sisters to be more closely related than are mothers to their daughters. He viewed this circumstance as favoring the formation of a sterile worker caste devoted to the care of sisters instead of daughters—and hence a major factor in the origin of eusociality. Haplodiploidy does not dictate the formation of sterile castes. Rather, it creates a bias in which other selection processes can successfully direct the evolution of species across the eusociality threshold.

Hamilton's proposition represents the most powerful form to date of the kin-selection hypothesis of the origin of social behavior. Other, competing explanations have been proposed, the most plausible being the queen-domination hypothesis of Alexander (1974) and Brothers and Michener (1974); but none has accounted for the details of hymenopteran social organization with comparable fidelity.

In the present chapter the mechanics of kin selection in Hymenoptera will be analyzed somewhat more closely than has been attempted in the past. Our purpose is to show that the process is not limited in its effects to caste formation but also has direct consequences for the evolution of caste ergonomics. We will also undertake to evaluate the strength and weaknesses of kin-selection theory in more detail than has been attempted in the past.

3.1. THE HAPLODIPLOID SYSTEM IN HYMENOPTERA

Sex in Hymenoptera is determined by a most elementary mechanism: females arise from fertilized eggs and are therefore diploid, while males arise from unfertilized eggs and are haploid. In breeding experiments with honeybees it has been shown that maleness is actually based on a sufficiently high level of homozygosity among certain genetic loci, a threshold which is automatically crossed when individuals are parthenogenetic in origin and haploid. Thus, it is possible to obtain diploid males by sufficient degrees of inbreeding within experimental stocks (Kerr, 1962).

The original adaptive significance of haplodiploidy remains moot. One likely hypothesis is that its main function was to permit females to adjust the sex ratio of eggs laid on prey objects according to the size of the objects and hence their suitability to support one sex or the other (Hamilton, 1967).

Whatever its original function, haplodiploidy has had a profound influence on the subsequent evolution of sociality in the Hymenoptera. During the nuptial flight the females mate with one or at most a few males. The spermatozoa are stored in her spermatheca; each can be paid out to fertilize an egg, or not, as the queen "chooses." Fertilized eggs develop into females—either workers or new queens—while unfertilized eggs become males. As a consequence, females inherit half of their genome from each parent and males inherit their entire genome from their mother. This circumstance produces asymmetries in the degrees of relationship, as noted by Trivers and Hare (1976); these are illustrated and explained in Figures 3.1 and 3.2.

The coefficient of relatedness, r_{ij}, between an individual i and a relative, j, can be defined as the average number of alleles in j which are identical by descent to alleles in i. Figure 3.2 shows why $r_{ij} \neq r_{ji}$ for the relationship between males and females. The numbers given in Figure 3.1 are based on three simplifying assumptions. (1) There is a negligible amount of inbreeding. Zero inbreeding is a reasonable approximation in many cases, because the majority of ant and termite species engage in large nuptial swarms drawn from hundreds or even thousands of colonies. Exceptional species of ants with small breeding populations are cited by Wilson (1963); their r_{ij}'s can be modified according to the procedures described by Hamilton (1972). (2) The queen is inseminated by a single male. This is a common but not universal trait; data are summarized by Wilson (1971), Hamilton (1972), and Michener (1974). (3) The colony is monogynous. The occurrence of monogyny and polygyny was described in Chapter Two. If any of the three premises are violated, the r_{ij}'s can be modified appropriately.

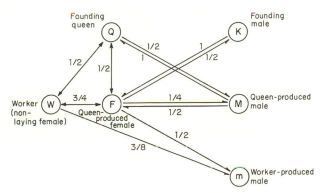

FIGURE 3.1. The degrees of relatedness, r_{ij}, within a colony of ants or other social hymenopterans. The queen-produced female F can be either a virgin queen or a worker (the category of worker is represented separately in the Figure).

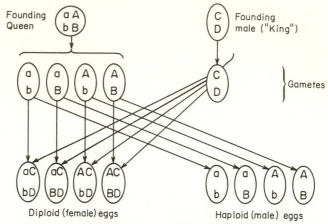

FIGURE 3.2. Schematic of the haplodiploid sex determination system, to be used with reference to Figure 3.1. Unfertilized eggs become males which carry on the average 1/2 of the queen's genome; that is, $r_{QM} = 1/2$. However, $r_{MQ} = 1$, since males receive all of their genome from their mother. Similarly, $r_{KF} = 1$, since the father donates all of his genome to his daughters, while $r_{FK} = 1/2$, since the daughter gets 1/2 of her genes from her mother ($r_{QF} = r_{FQ} = 1/2$). The key relationship for kin selection is $r_{FF} = 3/4$. This can be understood by focusing attention on any one of the individual daughters, say the one on the far left $abCD$. Her average relatedness to her sisters is $1/4[1 + 3/4 + 3/4 + 1/2] = 3/4 = r_{FF}$. Similarly $r_{FM} = 1/4[1/2 + 1/4 + 1/4 + 0] = 1/4$ for the same female.

Another way of arriving at $r_{FF} = 3/4$ is as follows. Focus on a gene, call it A, in female F_1. Denote by $Q \rightarrow A$ and $K \rightarrow A$ the event that the gene A in F_1 came from the queen and king, respectively. Since the queen is diploid, the probability that the A gene is in a sister, F_2, is then

$$\text{Prob } (A \text{ in } F_2/Q \rightarrow A) = 1/2$$

while the king is haploid, so

$$\text{Prob } (A \text{ in } F_2/K \rightarrow A) = 1$$

Since $\qquad \text{Prob } (Q \rightarrow A) = \text{Prob } (K \rightarrow A) = 1/2,$

$$\begin{aligned}
\text{Prob } (A \text{ in } F_2) &= \text{Prob } (A \text{ in } F_2/Q \rightarrow A) \cdot \text{Prob } (Q \rightarrow A) \\
&\quad + \text{Prob } (A \text{ in } F_2/K \rightarrow A) \cdot \text{Prob } (K \rightarrow A) \\
&= \tfrac{1}{2} \cdot \tfrac{1}{2} + 1 \cdot \tfrac{1}{2} = \tfrac{3}{4}
\end{aligned}$$

Thus, the probability that sisters, F_1 and F_2 share a common gene is just $r_{FF} = 3/4$.

3.2. THE INCLUSIVE FITNESS OF THE QUEEN AND WORKERS

Hamilton observed that since a female is more closely related to her sisters than to her own progeny ($r_{FF} = 3/4$ versus $r_{Fm} = 1/2$) she might project more of her genes into future generations by helping her mother raise sisters rather than by trying to become a mother herself. To make this notion more precise it is necessary to define the inclusive fitness of the queen and workers (Hamilton, 1964).

The queen can pass her genes to subsequent generations by way of three conduits coexisting in the same colony: (1) her daughters, related to her by $r_{QF} = 1/2$; (2) her sons, $r_{QM} = 1/2$; and (3) her grandsons, the sons of her daughter workers, $r_{Qm} = 1/4$. Let $\mathbf{N} = (F, M, m)$ denote the number of each reproductive type produced by the colony at the time of the nuptial flight. The efficiency with which the three agents can propagate the genes of the queen depends on the r_{ij}'s and the reproductive success each agent enjoys. For the moment the reproductive success will be represented merely as S_F, S_M, and S_m with reference to females, queen-produced males, and worker-produced males, respectively. An explicit expression for these terms will be derived later. Meanwhile, the *inclusive fitness of the queen*, V_Q, can be written as

$$V_Q = \sum_{j=1}^{3} r_{Qj} N_j S_j = \frac{1}{2} FS_F + \frac{1}{2} MS_M + \frac{1}{4} mS_m. \qquad (3.1)$$

Henceforth both classes of males are taken (on sound empirical grounds) as equal with reference to reproductive success, that is, $S_M = S_m$. Then, Equation (3.1) can be rewritten

$$V_Q = S_F \left[\frac{1}{2} F + \frac{1}{2} SM + \frac{1}{4} Sm \right]$$

$$= \frac{S_F}{2} \left[F + S \left(M + \frac{m}{2} \right) \right], \qquad (3.2)$$

79

where $S = S_M/S_F$ is defined as the *reproductive success ratio*. By the same reasoning the inclusive fitness of the workers is

$$V_W = \frac{3}{4} FS_F + \frac{1}{4} MS_M + \bar{r}mS_M, \qquad (3.3)$$

where $\bar{r} = 3/8$ for nonlaying workers and $\bar{r} = 1/2$ for a laying worker. Since in the majority of species most or all workers are nonlaying, $\bar{r} \cong 3/8$, and we can write:

$$V_W = \frac{S_F}{2} \left[\frac{3}{2} F + S \left(\frac{1}{2} M + \frac{3}{4} m \right) \right].$$

The first thing to notice is that $V_Q \neq V_W$. Thus, as deduced originally by Trivers and Hare (1976), there appears to be a fundamental conflict of interest between the queen and her daughter workers over the kinds and ratios of reproductives that should be produced by the colony.

3.3. REPRODUCTIVE SUCCESS DEPENDS ON THE COMPOSITION OF THE COMMUNITY

In order to proceed further it is necessary to develop an expression for the reproductive success of females as opposed to males. The reasoning leading to this step is as follows.

The chance that a particular male or female will successfully mate and contribute to the founding of a colony depends on the number of other reproductives competing in the population. In particular, the reproductive success of each sex in a given colony depends on the sex ratio in the entire breeding community. Consequently the inclusive fitness of each queen and worker cannot be assessed independently of the community dynamics. This connection creates a formidable problem unless we make some simplifying assumptions. The principal simplifying assumption we shall make here is that the population is at a steady state. Consider a community consisting of n colonies. For the time being we denote by F_i, M_i, and m_i the number of reproductive females, queen males, and worker males, respectively, in the ith colony. Therefore, the community composition is given by

$$\hat{F} = \sum_{i=1}^{n} F_i, \; \hat{M} = \sum_{i=1}^{n} M_i \text{ and } \hat{m} = \sum_{i=1}^{n} m_i.$$

(Circumflexes will denote community quantities.) The community sex ratio and male ratio is then given by $\hat{\sigma}(n) \triangleq (\hat{M} + \hat{m})/\hat{F}$ and $\hat{p}(n) \triangleq \hat{M}/(\hat{M} + \hat{m})$. The dependence on n, the community size, will be an important consideration in later modeling.

Next, let us give a precise definition of reproductive success. Consider males first, and define:

$S_M^{\tau} \triangleq$ the expected number of genes in generation τ which are identical by descent to a particular gene in a male in generation 0.

In other words: at generation 0 we "mark" a particular gene ("paint it red") and allow it to replicate and spread through the population. τ generations later we reexamine the population and count the number of "red" genes that are descended from that original "red" gene. We define the reproductive success of the marked male, M, as follows (see Wossy, 1978):

$$S_M \equiv \lim_{\tau \to \infty} S_M^{\tau}.$$

As noted, we shall assume that $S_M = S_m$, since males are probably indistinguishable with respect to mating success. With the above definitions we are now in a position to write a recurrence relation for S_M.

In a stationary community of n colonies the probability of a particular male founding a colony is $n/(\hat{M} + \hat{m})$. In each such colony founded there are, on the average, $\hat{F}/n$ females, $\hat{M}/n$ queen males, and $\hat{m}/n$ worker males. Each female carries a fraction r_{MF} of the founding male's genes and each male carries

81

$r_{MM} = 0$ or $r_{Mm} = 1/2$. Thus, we can write

$$S_M^{t+1} = \frac{n}{\hat{M} + \hat{m}} \left[\frac{\hat{F}}{n} \cdot 1 \cdot S_F^t + \frac{\hat{M}}{n} \cdot 0 \cdot S_M^t + \frac{\hat{m}}{n} \frac{1}{2} S_m \right]\Big|_t$$

$$= \left[\frac{1}{\hat{\sigma}} S_F^t + \frac{1 - \hat{p}}{2} S_M^t \right]\Big|_t . \tag{3.4}$$

The right-hand side of this recurrence relation connects the reproductive success in succeeding generations. However, since we are concerned only with the steady-state situation, we can set $S_M^{t+1} = S_M^t \triangleq S_M$ and solve for $S = S_M/S_F$, the relative reproductive success of males to females:

$$S(n) = \frac{S_M}{S_F} = \frac{2}{\hat{\sigma}(1 + \hat{p})} = \frac{F}{\hat{M} + \dfrac{\hat{m}}{2}}. \tag{3.5}$$

(This relation was derived independently by Benford, 1978, and Wossy, 1978). S is written $S(n)$ to note its dependence on the community size n, a relationship that will be established shortly. This equation says that the relative average success of males as opposed to reproductive females in transmitting genes is proportional to the ratio of reproductive females to males in the population of colonies as a whole, with the contribution of worker-produced males being diluted by half. This dilution is due to the fact that each female has two carriers of her genes in each generation—her sons and daughters—while each male can pass genes only to his daughters.

An identical line of reasoning leads to a recurrence relation for the reproductive success of females, S_F. However, it turns out that this equation is not independent of Equation (3.4) and hence leads to the same expression for S. In order to arrive at an independent equation we note that the total number of genes (or chromosomes) in the stationary population is constant at $2\hat{F} + \hat{M} + \hat{m}$. If at time zero we introduce a pair of marked ("red") chromosomes into each female, and a single marked ("blue") chromosome into each male, then t generations later we must have:

$$2\hat{F} + \hat{M} + \hat{m} = 2\hat{F}S_F^{(t)} + (\hat{M} + \hat{m})S_M^{(t)}. \qquad (3.6)$$

In particular, at $t = 1$ we can solve (3.6) and (3.5) for S_F and S_M separately:

$$S_M(n) = \left(\frac{1}{\hat{p} + 2}\right)\left(\frac{\hat{\sigma} + 2}{\hat{\sigma}}\right). \qquad (3.7)$$

$$S_F(n) = \left(\frac{\hat{p} + 1}{\hat{p} + 2}\right)\left(\frac{\hat{\sigma} + 2}{2}\right). \qquad (3.8)$$

However, it will turn out that in order to calculate the optimum inclusive fitnesses we only need S, the success ratio.

Equations (3.7) and (3.8) can now be substituted into Equations (3.3) and (3.4) to obtain the inclusive fitnesses of the queen and worker:

$$V_Q(F,M,m,\hat{p}(n),\hat{\sigma}(n))$$

$$= \left(\frac{\hat{p} + 1}{\hat{p} + 2}\right)\left(\frac{\hat{\sigma} + 2}{2}\right)\left\{\frac{1}{2}F + \left(\frac{1}{2}M + \frac{1}{4}m\right)\frac{2}{\hat{\sigma}(1 + \hat{p})}\right\}. \qquad (3.9)$$

$$V_W(F,M,m,\hat{p}(n),\hat{\sigma}(n))$$

$$= \left(\frac{\hat{p} + 1}{\hat{p} + 2}\right)\left(\frac{\hat{\sigma} + 2}{2}\right)\left\{\frac{3}{4}F + \left(\frac{1}{4}M + \frac{3}{8}m\right)\frac{2}{\hat{\sigma}(1 + \hat{p})}\right\}. \qquad (3.10)$$

3.4. THE CONFLICT BETWEEN THE QUEEN AND HER DAUGHTERS

We are now prepared to explore the connection between kin selection leading to preferred treatment among different sets of relatives, and colony-level selection, leading to improved ergonomic efficiency. It will be shown that the two do not always operate in concert.

Ergonomic constraints can be introduced into the discussion of inclusive fitness in the following way. Assume that each of the n colonies in the breeding population colonies has an

83

amount of energy, E, to allocate among the three reproductive categories: the potential queens, F, queen-produced males, M, and worker-produced males, m. Furthermore, we shall assume (on reasonable empirical grounds) that the average "manufacturing cost" for each type is fixed. These respective costs will be designated as C_F (calories/queen), C_M (calories/queen-produced male), and C_m (calories/worker-produced male). Thus, the ergonomic constraint imposed on each colony is

$$C_F F + C_M M + C_m m \leqslant E \qquad (3.11a)$$

or

$$\alpha_1 F + \alpha_2 M + \alpha_3 m \leqslant 1, \qquad (3.11b)$$

where α_i $(i = F,M,m)$ is the fraction of the total available energy allocated to reproductive type i. The situation is represented graphically in Figure 3.3. The ergonomic constraint, Equation (3.11), defines a simplex, Δ, which contains all feasible colony compositions, $\mathbf{N} = (F,M,m)$. As shown in the Figure a colony composition can also be specified by the quantities (E,σ,p).

If selection operates purely at the level of the colony, the determinants of fitness are the total numbers of the three kinds of reproductives weighted by their reproductive success:

$$
\begin{aligned}
V_C &= FS_F + MS_M + mS_m \\
&= S_F[F + S(M + m)].
\end{aligned}
\qquad (3.12)
$$

But this expression corresponds to neither the workers' nor the queens' inclusive fitness. Thus, we expect a clear conflict between the interests of the workers and queens on the one hand as opposed to that of their common enterprise, the colony. If colony-level selection is the final arbiter, then those colonies will be favored in which the queens and workers compromise so as to maximize V_C.

In order to find the colony composition $\mathbf{N} = (F,M,m)$ that maximizes each fitness separately, it is necessary to solve the

84

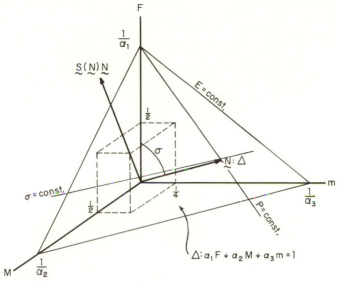

FIGURE 3.3. The colony composition can be represented by a vector $\mathbf{N} = (F,M,m)$. This vector is restricted by Equation (3.11) to lie on the simplex $\mathbf{\Delta}$, as shown. Points on $\mathbf{\Delta}$ can also be specified by the triple (E,σ,p). The queen's relatedness to her daughters, sons, and grandsons can be represented by a vector with components $(1/2,1/2,1/4)$. The queen can optimize her inclusive fitness by "selecting" a colony composition $\mathcal{N}$ such that the reproductive success vector $\mathbf{S}(\mathbf{N}) \cdot \mathbf{N}$ has as large a projection onto $\mathbf{r}_Q$ as possible. Since the workers' relatedness vector, $\mathbf{r}_W = (3/4,1/4,3/8)$, is not colinear with $\mathbf{r}_Q$ unless $M = 0$, the optima of the queen and workers generally will not coincide.

following optimization problem:

$$\underset{N}{\text{Max }} V_l = \sum_{i=1}^{3} r_{li} S_i \mathcal{N}_i \qquad l = Q, W, C, \qquad (3.13)$$

subject to

$$\alpha_i \mathcal{N}_i \leqslant 1; \qquad \mathbf{N} \geqslant 0. \qquad (3.14)$$

This is a straightforward problem in nonlinear programming. A geometric interpretation is given in Appendixes 3.1 and 3.2, and the reader is referred to Oster, Eshel and Cohen (1977) for further details. Here only the results will be discussed.

3.5. THE OPTIMUM INVESTMENT OF COLONY RESOURCES

Let us define the investment ratio by computing the total energy devoted to male versus reproductive female production:

$$\bar{R} = \frac{\text{Total cost of males}}{\text{Total cost of females}} = \frac{MC_M + mC_m}{FC_F}. \quad (3.15)$$

Using the quantities σ and p defined earlier we can write (3.15) as

$$\bar{R} = \sigma \bar{C}, \quad (3.16)$$

where

$$\bar{C} = \frac{pC_M + (1-p)C_m}{C_F}$$

is the average cost ratio of males to females. In Appendix 3.1 it is shown that the optimal ratio of investment R_l^* ($l = Q,W,C$) occurs at $p = 0$ or $p = 1$. From this fact R_l^* is computed in Appendix 3.2 by optimizing V_Q, V_W and V_C. The optimal values are given in Figure 3.4. If the queen has control over colony investment, she should prefer an equal allotment of resources to both males and females; in other words, $R_Q^* = 1$ regardless of whether her sons or grandsons constitute the entire male population. When the workers control the investment ratio, they prefer $R_W^* = 1/3$ if queen males are produced exclusively and $R_W^* = 1$ if worker males are produced exclusively; the

R^*	$m = 0$	$M = 0$
Q	1	1
W	1/3	1
CO	1	2

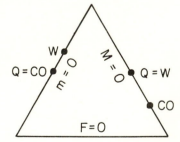

FIGURE 3.4. The optimal ratios of investment for the queen, workers, and colony: $R_l^* = \sigma \bar{C}$, $l = Q,W,C$. The corresponding points on the ergonomic simplex are shown to the right of the table.

latter case holds only when laying workers are greatly out-numbered by nonlaying workers (see Appendix 3.2). Thus, the workers agree with the queen only when they produce all of the males.

In cases where colony-level selection becomes paramount over kin selection, and males are all produced by workers ($M = 0$), then the colony investment ratio preferred by both the queen and her daughters is suboptimal; genes from colonies that maintain $R_C^* = 2$ will be preferentially represented in future generations. However, if queens produce the males ($m = 0$), the queen's preferred investment ratio agrees with the colony's best reproductive strategy, so that if workers dispute the queen's preference they imperil the fitness of the colony as a whole. This asymmetry would appear to constitute a strong imperative toward worker altruism, since the colony fitness is highest when the workers accede to the queen's preferred in-vestment ratio as well as relinquishing male production.

From the formulation of the optimization problem it can be shown that there are two areas of disagreement with respect to the optimization of fitness. The first involves queen-produced as opposed to worker-produced males,

$$0 \leqslant p \triangleq \frac{M}{M + m} \leqslant 1 \, ;$$

and the second is the sex ratio of reproductive forms produced by the colony,

$$0 \leqslant \sigma \triangleq \frac{M + m}{F} < \infty.$$

We have assumed that the "production costs" C_i (or the fractional production costs α_i) are constant, so that the ergo-nomic constraint is linear. In this case one can show that the optimum investment ratio for each party considered separately lies on the boundary of the ergonomic simplex, $p = 0$ or 1. Furthermore, even when the workers' and queen's fitnesses are considered jointly, that is, in the conflict situation, the competitive equilibria are also on the boundaries $p = 0,1$. In order to define precisely what we mean by a "competitive equilibrium" we must introduce the notion of "vector optimization," a concept that will reappear frequently in later chapters when we treat ergonomic optimization. A brief discussion of this concept is presented in Appendix 3.3.

3.6. WHAT ARE THE OPTIMUM INVESTMENT RATIOS?

The result of optimizing the inclusive fitnesses, V_l ($l = Q,W,C$) is that either queen-produced males (M) *or* worker-produced males (m) should be produced, never both simultaneously. The decision of which is more advantageous depends on the following cost-relatedness inequality: if

$$\frac{C_m}{C_M} \geqslant \frac{\alpha_3}{\alpha_2} \cdot \frac{r_{13}}{r_{12}}, \qquad l = Q,W,C \qquad (3.17)$$

then $m = 0$ is a necessary condition to maximize V_l ($l = Q,W,C$). The relation can be expressed in words as follows. When workers are in control they should tolerate the production of brothers (M's) by their mother rather than nephews (m's) by their sisters only if the cost of producing nephews relative to the cost of producing brothers exceeds the degree of their relatedness to nephews relative to their relatedness to brothers (in many cases, this last value will be $[(3/8) \div (1/4) = 1.5]$). In general, inequality (3.17) implies that all of the males of a colony will come from either the queen or from the workers, i.e., $p = 0$ or 1, if one or the other party controls the male production.

3.7. WHAT ARE THE LIMITATIONS OF THE KIN-SELECTION MODEL?

The empirical data gathered by Trivers and Hare (1976) for ants show that the ratio of investment is typically lower than 1 and frequently in the vicinity of 1/3. Also, as noted earlier, in most ant species for which information is available the queen produces all the males ($m = 0$). Thus, we appear to have the general picture within the ants of workers that control colony reproduction and tip the ratio of investment to their own advantage. The simplest explanation would entail the following processes:

(1) workers so outnumber the queen that they have direct physical control over the brood almost all the time, a fact that has been confirmed by behavioral observations;

(2) the cost ratio C_m/C_M is indeed significantly higher than the relatedness ratio r_{13}/r_{12}, see Equation (3.17), giving the workers greater inclusive fitness when they rear brothers instead of nephews; and

(3) colony-level selection does not affect the investment ratio as strongly as kin selection.

This relatively straightforward formulation is complicated, however, by technical questions concerning the data gathered by Trivers and Hare. These difficulties, which have been stressed by Alexander and Sherman (1977), are too intricate and detailed to be easily summarized here. Even if the objections prove invalid or trivial, the Trivers-Hare data as published do contain a wide scatter that must be taken into account. A plot of $\log[C_F F]$ against $\log[C_M M + C_m m]$ gives a principal correlation axis corresponding to $\bar{R} = 1/2.9$ with 95 percent confidence limits of (2.0-4.9) (R. Colwell, personal communication). Alexander and Sherman argue that the lowered investment ratio cannot be interpreted as any particular number such as 1/3. They believe that the data can equally well be explained as the outcome of competition among related males in populations of small effective size, so that mother queens and workers would find it profitable to invest in a smaller number of males as opposed to females. This alternative hypothesis is not well supported by information on the ant species, many of which engage in huge nuptial flights and therefore are seldom put in the situation where related males compete to any significant degree.

At this point it must be admitted that the lowered investment ratio discovered by Trivers and Hare, while consistent with the theory, cannot be regarded as decisive evidence for kin selection as some previous reviewers have concluded (for example, Wilson, 1975a). More investment data are needed, and the species from which the data are collected should be partitioned according to the critical factors of degree of inbreeding and the provenance of the males. Even with these improvements, it will be difficult to erase all ambiguity. The theoretical optimal solution of the ratio of investment turns out to be unusually "flat." If the number of colonies in the breeding community is moderately large, say $n > 10$, the fitness penalty for deviating substantially from the optimum point will be quite small. Therefore, it is possible that selection for sex ratios is relatively weak, and the sex ratio data could easily show a wide scatter due to random influences. This contribution to the variance might make the kin-selection hypothesis difficult to test on the basis of investment data alone. Other, stronger confirmations of kin-selection have been provided by the details of the much less altruistic and cooperative behavior of males within hymenopteran societies (Hamilton, 1964; Wilson, 1971; Wilson and Hölldobler, 1978), and still other, ingenious tests will surely be invented in the future.

In extending the kin-selection model we have made a series of assumptions, some of the more questionable of which are listed in Table 3.1.

Assumptions (1) and (2) were required to derive an expression for the reproductive success ratio employed in our definition of inclusive fitnesses. If stochastic perturbations or complex demographic dynamics are strong ongoing features of normal community growth, then it can be shown that "chaotic" polymorphisms are sustainable (Auslander, Guckenheimer and Oster, 1977). In such cases natural selection cannot optimize inclusive fitness, or anything else. The optimization problem formulated in Equation (3.13) is a static one and does not ac-

TABLE 3.1. Biological assumptions made to derive the optimal ratios of reproductive investment (see Figure 3.4).

---

(1) The population of colonies is at a steady state.

(2) There is unrestricted competition for mates over the entire population of colonies (see discussion in Hamilton, 1967, and Alexander and Sherman, 1977).

(3) All colonies are ergonomically identical; that is $\bar{M} = nM$ and $\sum \alpha_i N_i = E$ for all colonies.

(4) The community of colonies is panmictic. For populations of colonies with small effective breeding sizes, the r_i's must be multiplied by a function of the inbreeding coefficient (see Hamilton, 1972).

(5) The number of laying workers is much less than the total number of workers and brood care is indiscriminate, so that the average relatedness of a given worker to the male brood is close to 3/8 rather than to 1/2, the value that would exist if each worker nurtured her own offspring.

---

count for the colony and community dynamics, which should be regarded as additional constraints. That is, a more precise formulation is

$$\text{Max } V_l(\mathbf{N}, \hat{\sigma}, \hat{p}),$$

subject to

$$\int \sum \alpha_i N_i \, dt \leqslant \text{const.}, \qquad N \geqslant 0,$$

and

$$\dot{\mathbf{N}} = \mathbf{F}(\mathbf{N}, \hat{\sigma}, \hat{p})$$

$$\hat{\mathbf{N}} = \sum_{k=1}^{n} N_k, \qquad N_k > 0, \qquad k = 1, 2, \ldots, n.$$

The last two equations, viewed as dynamic constraints, could radically alter the nature of the optimum fitnesses computed on the basis of the time-averaged static model (see Chapter Eight). In particular, they would increase the variance of the investment ratios in ways difficult to analyze.

Assumption (3), that all colonies are identical, is equally suspect. Cohen and Eshel (1976) and D. S. Wilson (1975) have demonstrated that this situation can be unstable under certain conditions where random dispersal of alates is taken into account. We will discuss this circumstance in the next section. It is possible that no reproductive system exists which is optimal for every colony. Under such circumstances one could expect a mixture of investment strategies through the population of colonies, with some colonies specializing in males, others in females, and others in mixtures. Only the sex ratio *distribution* over the population of colonies need be stable.

Inbreeding is one possible outcome of altered spatial distribution, as noted previously. Inbreeding increases the relatedness of the queen to her daughters by a factor of $1 \leqslant (1 + 3F)/(1 + F) \leqslant 2$, so that the discrepancy between $R_{\widetilde{Q}}^{*}$ and R_{W}^{*} is diminished.

Assumption (2) must be untrue in a great many instances. In species that found colonies by swarming, such as army ants and honeybees, the sex ratio is heavily biased toward males. This difference is apparently associated with the disproportionately low risk faced by colony-founding queens (which are assisted by thousands of workers) as compared with the risk taken by males who must compete for mates during the nuptial flights. The apparent heavy bias toward males in fission-reproducing colonies can be brought back more in line with kin-selection theory if we include along with the female costs the investment in the worker swarm which accompanies the queen, i.e.,

$$R = \frac{C_M M + C_m m}{C_F F + C_W W},$$

where $C_W W$ is the total cost of the nuptial swarm (Macevicz, 1978).

Finally, nowhere in our analysis were gene frequencies calculated—our treatment was purely phenomenological referring to the degree of relatedness parameters, r_l. A detailed genetic treatment is possible in principle, but difficult to carry out in practice. An elementary version is presented by Oster, Eshel, and Cohen (1977).

3.8. HOW CAN THE INTERMEDIATE CASE $0 < p < 1$ BE CONSISTENT WITH THE KIN-SELECTION MODEL?

At least one prediction of the theory of investment ratios can be tested more directly: Have species of social Hymenoptera in fact opted for all-queen or all-worker production of males, rather than mixtures? The sample of species for which this information exists is not large enough to make a secure generalization, but it does suggest that the prediction is usually met. In many ant species belonging to a phylogenetically wide array of genera (*Eciton, Monomorium, Pheidole, Solenopsis*) the workers completely lack ovaries; hence it is certain that all males come from the queens. Workers of the harvester ant *Pogonomyrmex badius* lay eggs, but they appear to be invariably trophic in nature and are promptly eaten by the larvae or queen. By distinguishing allozymes electrophoretically, Crozier (1974) was able to show that all of the males of the ant *Aphaenogaster rudis* come from the queen, in spite of the fact that the workers have well-developed ovaries. When honeybee queens are in residence they are responsible for virtually all of the males produced by the colony; they lay unfertilized eggs into special drone cells.

We know of no case in which the workers are the *exclusive* parents of males when healthy queens are present, but mixed parentage appears to be relatively common. According to R. Zucchi (in Michener, 1974) workers of the South American bumblebee *Bombus atratus* lay 90 percent of the male-destined eggs. In some of the North American species of *Bombus*, the queens produce at least 50 percent; it is possible that they are normally the parents of all the males during the main period of colony growth, with workers supplementing them or taking over entirely when they grow feeble or die near the end (C. Plowright, personal communication). Mixed parentage appears widespread in the eusocial halictine and meliponine bees (review by Michener, 1974). In the meliponine *Trigona postica* workers often lay an unfertilized egg in a cell next to the somewhat smaller fertilized egg of the queen. The larger and more active male larva then bites and kills the female larva, so that the cell eventually gives rise to a single male (Beig, 1972).

Exclusive worker parentage has been considered as a possibility in a small minority of queenright colonies of wasps of the genus *Vespula* by Montagner (1963, 1966) and Spradbery (1973). On the other hand, Montagner established that in some of the *Vespula* colonies queens are the exclusive parents of males, while in about half the colonies workers contributed to approximately 75 percent of the male parentage. Mixed male parentage has also been reported in colonies of *Polistes fadwigae* by Yamanaka (1928). Brian (1968) has determined that parentage in *Myrmica* colonies is mixed, with most males coming from workers.

This striking variation of male provenance among species of social Hymenoptera is a subject needing future study. Explanations other than the kin-selection hypothesis offered here are possible. For the moment we will pursue the implications of the reported cases of simultaneous mixed parentage with reference to the kin-selection model. Three explanations can be considered.

(1) *Mixed Parentage Due to Varying Costs*

First, costs (C_i) could be a compound of energy requirements that varies over the colony life cycle. Until now the only element considered has been the energy required to rear a given type of individual $(F, M,$ or $m)$ to adulthood. But there are likely to be other, "hidden" costs. If laying workers cease to forage, the potential energy lost (C_{LOST}) plus the metabolic costs of the nonforaging workers (C_{MET}) must be added to the expense of rearing males:

$$\hat{C}_3 = C_3 + C_{LOST} + C_{MET}$$

If the foraging efficiency of laying workers is significantly less than that of non-laying workers, then the energy constraint becomes a function of p as shown in Figure 3.5. In this case the inclusive fitness optimization will generally yield an intermediate value of $0 < p < 1$. It is also very likely that the costs vary over the course of the colony cycle. For example, a large colony might well operate at an energetic surplus because of the economics of scale discussed in Chapter Two, and it could therefore easily afford to subsidize a number of workers disposed toward egg laying.

(2) *Mixed Parentage Due to Irregular Social Control*

A second potential explanation of the mixed parentage of males within a colony entails the control of colony investment. In most of the eusocial hymenopteran species in which mixed

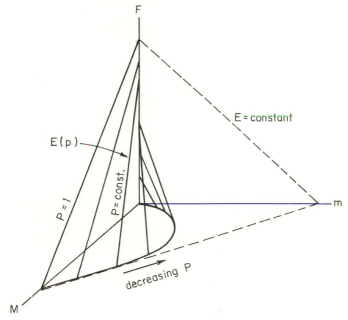

FIGURE 3.5. The ergonomic constraint surface will generally be non-linear if the energy available is a function of the proportion of active foraging (nonlaying) workers.

male parentage occurs, particularly the bumblebees and paper wasps, the queen directly inhibits worker egg-laying through aggressive behavior. Frequently, as the season progresses the queen's power begins to wane. This is due either to senescence in the annual colonies, as in the case of bumblebees, or because the colony grows too large for the queen to police. In either event one or a few workers may commence to lay eggs of their own, while taking on some aspects of regal behavior and dominating other workers. The conflict is sometimes exacerbated by reciprocal egg-eating. The net result is that by the time of the nuptial flight the colony's output consists of a mixture of worker-produced and queen-produced males. The queen's failure to enforce her R^* for the entire reproductive period means that the net colony fitness is a sequential combination of V_Q and V_W, as shown in Figure 3.6. A similar weakening of the queen's

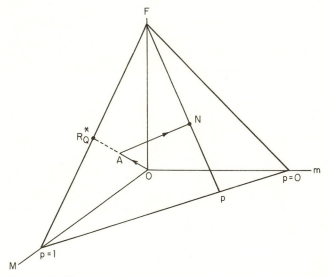

FIGURE 3.6. The colony composition, **N**, at nuptial flight time may consist of a mixed brood of the queen's sons and the workers' sons. This could occur if the queen controlled male production early in the reproductive phase, **O** to **A**, and the workers started laying eggs in **A** to **N**.

control can occur when the nest structure becomes very diffuse, a common situation in such ant genera as *Odontomachus*, *Formica*, and *Oecophylla*.

(3) *Mixed Male Parentage Due to Group Selection*

Several authors have formulated models of group selection in which altruistic individuals can be subsidized in a population (Cohen and Eshel, 1976; D. S. Wilson, 1975; Mirmirani and Oster, 1978). These models depend on two effects for their operation: (a) altruists confer a net benefit on their group so that the group as a whole fares better, even though within each group altruists are at a selective disadvantage; (b) periodically the system is reassorted into new groups by some random process comparable to seed dispersal in annual plants. An analysis of how competitive equilibrium can be achieved in such a situation has been presented by Mirmirani and Oster (1978). It can be shown that in annually reproducing colonies this effect can stabilize a polymorphism between altruistic nonlaying workers and nonaltruistic laying workers, providing that laying workers penalize the colony energetically as discussed in (1) above.

3.9. THE RELATION BETWEEN REPRODUCTIVE FITNESS, ERGONOMIC EFFICIENCY, AND CASTE EVOLUTION

Regardless of the outcome of the queen-worker conflict, there should always be a strong selective force toward increased colony efficiency. The number of reproductives ($N = F + M + m$) produced by the colony at the onset of the reproductive stage is a monotonically increasing function of the number of ergonomically productive workers accumulated during colony growth in the ergonomic stage (Wilson, 1971, 1974a; Spradbery, 1973). From the point of view of the colony, as large a worker force as possible should be generated to maximize the subsequent manufacture of reproductives. Because the colony growth rate is

97

generally a function of the number of workers and not of the number of egg-layers, the colony will yield more reproductives by nuptial flight time if the workers forego reproduction in favor of the queen. No problem exists if the queen is in control, since she can produce all the males ($m = 0$), and as a result $V_Q = V_C$ (see Figure 3.4). In that case, colony-level selection in social hymenopteran populations is virtually the same as selection on queens.

On the other hand, if workers control the colony investment, the usual circumstance in the social Hymenoptera, and if male production is left in the hands of the queen, then $V_W \neq V_Q = V_C$. Hence a component of worker-inclusive fitness can exist that is orthogonal to ergonomic efficiency. That efficiency is still the coin of the realm can be verified, however, by reference to the graphical analysis presented in Figure 3.7. Here contours of constant V_Q and V_W are superimposed on the feasible set of compositions of the reproductives produced by a colony (M,F). The maxima of V_Q and V_W are located at the points W and Q respectively on the ergonomic constraint line $E = C_F F + C_M M$. The line segment (W-Q) on E is the region of dispute. That is, points on the energy frontier E either to the left of W or to the right of Q are disadvantageous to both workers and queens, and both castes would profit by moving the investment along E until the segment (W-Q) is reached. Once onto (W-Q), however, neither party can gain fitness without the other losing. Although the queen has additional selective leverage due to the fact that her fitness is colinear with that of the colony, the workers perform the task of brood rearing and therefore are more likely to control the sexual composition of the reproductive brood. Thus, the relative "power" of the queen is pitted against that of her daughters in the region (W-Q). In the parlance of game theory, points along (W-Q) are competitive ("Nash") equilibria or "Evolutionarily Stable Strategies" (see Maynard Smith [1974]). A more complete discussion is given in Appendix 3.3 and in Oster, Eshel, and Cohen (1977).

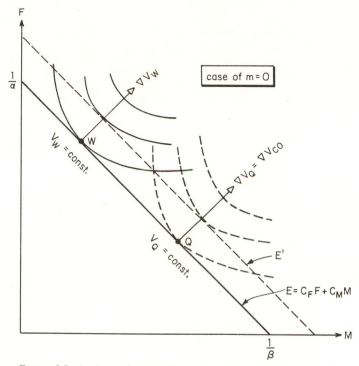

FIGURE 3.7. A colony of social hymenopterans has a limited amount of energy E to invest when creating new reproductive forms, which consist of the new virgin queens F and males M. Because of asymmetries in the genetic relationships that stem from haplodiploid sex determination, the maximal inclusive fitnesses of the workers V_W and of the mother queen V_Q will differ when the new males are generated by the mother queen. In spite of the disagreement over investment in new queens versus males along the Nash equilibrium segment W-Q, *any* increase in ergonomic efficiency of the colony as a whole will increase the inclusive fitness of all parties because it shifts the energy line outward to E'.

If the workers control the investment, it appears that colony fitness (V_C) must be sacrificed to some degree. However, the colony fitness can actually be raised if, as a concomitant of the control, the workers increase ergonomic efficiency, extending the ergonomic constraint from E to E' as depicted in Figure 3.7. One of the avenues to such an improvement is the evolution of

99

a worker subcaste system. Natural selection will favor the evolution of specialized castes because any enhancement in ergonomic efficiency that results will be translated directly into reproductive fitness.

The situation can also be viewed from the perspective of our discussion of colony reliability in Chapter One. Recall that N_j in the inclusive-fitness formula refers to the *expected* number of alates of type j produced at nuptial flight time. That is, N_j should actually be written as PN_j, where P is the colony reliability, i.e., the probability of surviving to the reproductive phase. P is surely a monotonically increasing function of the number of nonlaying workers, especially soldiers, produced during the ergonomic phase. Thus, the structure of the caste system enters in a direct way into the expression for inclusive fitness. We shall model this connection explicitly in Chapter Six.

3.10. INDIVIDUAL-LEVEL SELECTION VERSUS CASTE MULTIPLICATION

Another aspect of the relation between reproduction competition within the colony and caste evolution is the inhibitory effect the first can exercise on the second. To develop into an extreme caste is to surrender reproductive potential. Workers of the social Hymenoptera are by and large not wholly sterile; those of many species have functional ovaries, produce trophic eggs, and possess at least the capacity to lay unfertilized eggs that ordinarily yield sons. Is it really to the advantage of hymenopteran workers to give up this opportunity? As we have seen, the answer is frequently no: under a variety of circumstances where the workers control the ratios of investment in reproductive males and females, it is advantageous for individuals to rear their own sons and nephews, while continuing to rear the fertilized eggs received from their mothers, which yield sisters. R. L. Trivers (personal communication) has suggested to us that this advantage of selfish behavior might serve

100

as a constraint on the evolution of castes (see also discussion in Wilson, 1976a). We believe that this may well be the case, that the number of worker subcastes and the degree of their specialization does indeed represent the balance struck between individual-level selection on the one hand, which tends to make workers ever more queenlike and hence convergent to each other, and colony-level selection on the other hand, which tends to proliferate sterile castes. It is even remotely possible that monomorphism represents the prevalence of individual-level selection, rather than an adaptation to the environment in which generalists, all resembling one another, are the optimum caste for the colony as a whole. Where then is the balance struck—how powerful a constraint is individual-level selection, in comparison with the other, ergonomic constraints?

Testing for at least a minimal role of individual-level selection in the limitation of caste evolution is not impossible. There should be a positive correlation between monomorphism and the possession of ovaries by the worker caste. Note that to fulfill the condition of reproductive potency the workers do not have to possess functional ovaries capable of producing viable eggs all during their lives. The condition would be met even if workers oviposited only when the queen is removed. This is in fact the condition in *Aphaenogaster rudis* (Crozier, 1974). Our first assessment, summarized in Table 3.2, indicates that the correlation does exist. The data shown are very incomplete, because the presence or absence of ovaries has never been systematically investigated in ants. However, even prior to extending the search, the available information suggests to us the existence of a significant trend. The difference between the ratios A/C ($=25/4$) and B/D ($=2/4$) is highly significant. It is even more significant when one more closely examines category B, which together with C contains the genera that depart from the expected association. Both of the B taxa, *Monomorium* (*sensu stricto*) and *Solenopsis* subgenus *Diplorhoptrum*, are char-

TABLE 3.2. The association between the degree of polymorphism and the presence or absence of worker ovaries in ants. The data given are the numbers of ant genera and subgenera in each of the four possible categories.

	Workers with ovaries	Workers lacking ovaries
Monomorphic or weakly polymorphic	A 25	B 2
Strongly polymorphic, including completely dimorphic	C 4	D 4

acterized by the presence of dwarf workers. In the case of *Diplorhoptrum*, the "thief ants," exceptionally small size provides the advantage of permitting the workers to forage into narrow spaces and to rob brood from the nests of other kinds of ants. We suggest that this small size may carry with it the price of the loss of ovaries, hence the deviation from the predicted association.

We also suggest that the Ponerinae might be viewed in a different light. The fact that their species are almost all monomorphic has previously been interpreted as simply a reflection of their more primitive phylogenetic state. In many other anatomical features they are indeed less advanced than most other ants. But the Ponerinae have existed since at least Eocene times and their history may go back into Cretaceous times, or over 70 million years (Dlussky, 1975). It is at least remotely possible that individual-level selection has simply held sway, with workers preserving their reproductive option. In fact, ponerine workers generally possess ovaries.

This leads to a key question: under what ecological conditions would the balance be tipped toward individual-level selection of such relative intensity as to result in monomorphism? Here are some possibilities:

(1) Trivers and Hare (1976) have predicted that workers should be reproductively more active in polygynous species, that is, species in which colonies contain multiple queens. The basis of this inference is that when multiple queens are present, workers have less to lose by entering into conflict with individual queens during disputes over who will lay male-destined eggs. When only a single queen is present, she loses little inclusive fitness by injuring a single worker, but the worker has a great deal to lose. This asymmetry is less pronounced when multiple queens are present. The Trivers-Hare prediction has not yet been tested.

Meanwhile, is there also an association between polygyny and monomorphism, as predicted by extending this argument? The available information is too sparse to decide. The association, or lack of it, might be the subject of systematic data gathering in the future.

(2) If worker oviposition is indeed conducted in the face of opposition from the mother queen, we should expect this control to be least effective in the largest colonies, where the queen is least able to monitor the activities of individual workers. Thus, all other things being equal, species with the largest colonies should be more frequently monomorphic than those with small to medium colonies. Unfortunately for the hypothesis, the reverse is true. However, there is a competing explanation for this relation: species with large colonies are also those with the most complex adaptations, requiring an array of castes, for example, leafcutters and army ants. They may also be those with the greatest tempo and turnover of workers, requiring a finer division of labor.

SUMMARY

The haplodiploid mode of sex inheritance characteristic of the Hymenoptera creates asymmetries in the degrees of relatedness between the queens, workers, and males. Because of the asymmetries, the optimal investments on the part of the queen, workers, and colony as a whole will not coincide under many conditions, and conflict between the queen and workers over the production of males is a possibility in two areas:

(1) In order to maximize the inclusive fitness of a given agent l (l = queen, worker, or colony), all the males should be produced by the queen if $C_m/C_M > r_{lm}/r_{lM}$, where C_m/C_M is the ratio of the energetic cost in rearing individual worker-produced males as opposed to the cost of rearing queen-produced males, and r_{lm}/r_{lM} is the ratio of relatedness of the agent l to worker- and queen-produced males; see Equation (3.17). If this inequality is not met with reference to workers, their inclusive fitness will be maximized by rearing their own sons and nephews, and hence bring the workers into potential conflict with the queen. In either case, the optimum strategy should always be to give exclusive parentage either to the queen or to the workers, and not to permit mixed parentage.

(2) The optimum ratios of investment in virgin queens versus males are different from the viewpoint of the queen, worker, and colony and according to whether the males originate exclusively from the queen or from the workers; see Figure 3.4.

Conflict is contingent on the inevitable limitation of the energy available for investment and maintenance of the reproductive forms. Regardless of the nature and intensity of the conflict, any improvement in ergonomic efficiency will increase the inclusive fitness of the queen, the workers, and the colony as a whole and therefore should be invariably favored by natural selection (Figure 3.7). One of the principal modes of enhancing efficiency is caste formation and division of labor which can be translated into increased reproductive fitness for all parties. Yet the individual advantage gained by workers from raising sons and nephews does have the potential of inhibiting caste proliferation, since to become too specialized for certain tasks is to surrender the ability to lay eggs or to rear larvae. Some indirect evidence suggests that the inhibition has occurred in evolution, but its magnitude cannot be estimated at this time.

APPENDIX 3.1

The optimum inclusive fitness for each party, V_l^* ($l = Q, W, C$) is the solution to the following nonlinear programming problem:

$$V_l^* \triangleq \text{Max } S_F[r_{l1}N_1 + r_{l2}N_2 S + r_{l3}N_3 S], \qquad (3.18)$$

subject to the constraints

$$\sum_{i=1}^{3} \alpha_i N_i = 1, \qquad N_i \geqslant 0. \qquad (3.19)$$

We can interpret this geometrically by introducing the following vector notation: $\mathbf{N} = (F, M, m)^T$, $\mathbf{r}_l = (r_{l1}, r_{l2}, r_{l3})$ (superscript T denotes the matrix transpose),

$$\mathbf{S} = \begin{bmatrix} S_1(\hat{\mathbf{N}}) & 0 & 0 \\ 0 & S_2(\hat{\mathbf{N}}) & 0 \\ 0 & 0 & S_3(\hat{\mathbf{N}}) \end{bmatrix}$$

Then Equation (3.18) can be written:

$$\text{Max}_{\mathbf{N}} [\mathbf{r}^T \mathbf{S} \mathbf{N}] = V^*, \qquad (3.20)$$

subject to

$$\boldsymbol{\alpha}^T \mathbf{N} = 1, \qquad \mathbf{N} > 0. \qquad (3.21)$$

The geometric picture is the following: agent $l(= Q, W, C)$ selects a colony composition vector $\mathbf{N}$ on the ergonomic constraint surface with the following property. When $\mathbf{N}$ is scaled by the reproductive success matrix (which depends on the community composition $\hat{\mathbf{N}}$, and therefore on $\mathbf{N}$ itself) the resulting vector, $\mathbf{S}(\hat{\mathbf{N}}) \cdot \mathbf{N}$, has as large a projection as possible onto the agent's relatedness vector, $\mathbf{r}_l$. The situation with regard to the queen, for example, is shown in Figure 3.8.

105

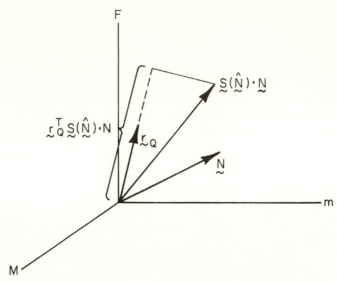

FIGURE 3.8. The preferred colony composition vector **N** of a particular agent (l = queen, worker or colony as a whole) depends on the composition of the entire population of colonies $\hat{\mathbf{N}}$. This diagram applies to the queen.

Equations (3.18) and (3.19) can be solved directly by Lagrange Multipliers (see Intriligator, 1971). The result is:

$$\mathcal{N}_1 = \frac{1}{\alpha_1} \left\{ \frac{2(\alpha_2 r_3 - r_2\alpha_3) + r_1(\alpha_2 - 2\alpha_3)}{2n(\alpha_2 r_3 - \alpha_3 r_2) + r_1(\alpha_2 - 2\alpha_3)} \right\} \triangleq \frac{1}{\alpha_1} \frac{K + Lr_1}{nK + Lr_1}$$

$$\mathcal{N}_2 = \frac{1}{L} \left\{ \frac{nK - 2r_3 L}{nK + r_1 L} \right\}$$

$$\mathcal{N}_3 = \frac{2}{L} \left\{ \frac{-nK + r_2 L}{nK + r_1 L} \right\}$$

where n is the number of colonies in the community that enters into the definition of S:

$$S(n) = \sum_{k=1}^{n} F_k \bigg/ \left[\sum_{k=1}^{n} M_k + \frac{1}{2} \sum_{k=1}^{n} m_k \right] \triangleq \frac{\hat{F}}{M + \frac{\hat{m}}{2}}.$$

The significance of computing the optimum fitness *before* letting $n \to \infty$ is explained in Appendix 3.2. The optimum sex ratio is

$$\sigma = \frac{N_1}{N_2 + N_3} = \frac{1}{\alpha_1} \frac{K + r_1 L}{2(r_2 - r_3) - n\dfrac{K}{L}}.$$

Now, it is clear that $L < 0$, since worker males are always more expensive than queen males. Furthermore, $K > 0$ only when $\alpha_2 > 2\alpha_3$ (for V_Q) and

$$\alpha_2 > \frac{2}{3}\alpha_3 \text{ (for } V_W).$$

Therefore, we can conclude that when either the queen or the workers control the brood proportions, the maximum values, V_Q^* and V_W^*, fall on the boundary of the constraint simplex $\Delta : \boldsymbol{\alpha}^T \mathbf{N} = 1$. In fact, we find that the maxima lie on the line $M = 0$ or $m = 0$. This can be easily seen by using (3.19) to eliminate N_3 from (3.18), obtaining:

$$V_l = S_F \left\{ N_1 \left(r_{l1} - S r_{l3} \frac{\alpha_1}{\alpha_3} \right) + S N_2 \left(r_{l2} - r_{l3} \frac{\alpha_2}{\alpha_3} \right) \right\}.$$

Thus, N_2 adds a positive contribution to V_l only if

$$\frac{r_{l2}}{r_{l3}} > \frac{\alpha_2}{\alpha_3}.$$

Thus, either $m = 0$ or $M = 0$ depending on whether the cost ratio of M's to m's is greater or less than the relatedness ratio to agent $l = Q,W,C$. Therefore, we can restrict our attention to the boundaries $(M = 0, m = 0)$ of the energy simplex in the search for the maximum inclusive fitness. This calculation is carried out in Appendix 3.2.

APPENDIX 3.2

Assuming that either $M = 0$ or $m = 0$, the optimization problem is:

$$V_l^* \triangleq \text{Max}\{S_F[r_1 F + r_2 S M]\}, \qquad l = Q,W,C, \quad (3.22)$$

subject to

$$\alpha F + \beta M = 1, \tag{3.23}$$

and where

$$S = \sum_{k=1}^{n} F_k \bigg/ v \sum_{k=1}^{n} M_k \begin{cases} v = 1 & \text{if} \quad m = 0 \\ v = \dfrac{1}{2} & \text{if} \quad M = 0. \end{cases} \tag{3.24}$$

$$\beta \to \alpha$$

Substituting Equation (3.23) into Equation (3.22):

$$V_l = S_F \left[\frac{r_1}{\alpha} + M_k \left(r_2 S - r_1 \frac{\beta}{\alpha} \right) \right]. \tag{3.25}$$

Then, $dV_2/dM_k = 0$ yields for the optimum, M^*:

$$0 = r_2 S - r_1 \frac{\beta}{\alpha} + r_2 M_k^* \frac{dS}{dM_k}. \tag{3.26}$$

To evaluate dS/dM_k, substitute (3.23) into (3.24):

$$S = \frac{1}{v\alpha} \frac{n - \beta \sum M_k}{\sum M_k} = \frac{1}{\alpha v} \frac{n}{M} - \frac{\beta}{\alpha v}. \tag{3.27}$$

Then

$$\frac{dS}{dM_k} = \frac{n}{\alpha v \hat{M}^2}, \tag{3.28}$$

(since $d\hat{M}/dM_k) = 1$. Then, from Equation (3.26):

$$0 = \frac{r_2 n}{vM} - \frac{\beta r_2}{v} - r_1 \beta + r_2 M_k^* \left(-\frac{n}{vM^2} \right) \tag{3.29}$$

Differentiating with respect to any colony's males gives the same answer, so that all n colonies are identical, i.e.,

$$M = \sum_{k=1}^{n} M_k = nM.$$

Then Equation (3.29) becomes

$$0 = \frac{r_2}{vM^*} - \frac{r_2\beta}{v} - r_1\beta - \frac{r_2}{vnM^*} \tag{3.30}$$

or

$$M^* = \frac{1}{\beta}\frac{r_2}{(vr_1 + r_2)}\left(1 - \frac{1}{n}\right). \tag{3.31}$$

In a large community, $n \to \infty$, the optimum is:

$$M^* = \frac{1}{\beta}\frac{r_2}{(vr_1 + r_2)}. \tag{3.32}$$

It is important to notice that the large community limit $n \to \infty$ cannot be taken until after the optimum $dV_2/dM = 0$ has been computed. This is because

$$\frac{d}{dM}\lim_{M \to \infty} S(\hat{M}) \neq \lim_{M \to \infty}\frac{d}{dM}S(\hat{M}), \tag{3.33}$$

since the series $\sum_{k=1}^{n} M_k$ does not converge (see Marsden, 1974, p. 108).

Using (3.31) in (3.23) the females are

$$F^* = \frac{1}{\alpha}\frac{vnr_1 + r_2}{n(vr_1 + r_2)} \tag{3.34}$$

and

$$\lim_{n \to \infty} F^* = \frac{1}{\alpha}\frac{vr_1}{(vr_1 + r_2)}. \tag{3.35}$$

The optimum sex ratio is:

$$\sigma^* = \frac{M}{F} = \frac{\alpha}{\beta}\frac{r_2(n - 1)}{r_2 + vnr_1} \xrightarrow{n \to \infty} \frac{\alpha}{\beta}\frac{r_2}{vr_1}. \tag{3.36}$$

The optimum ratio of investment, R, is:

$$R^* = \sigma \cdot \frac{\beta}{\alpha} = \sigma C, \tag{3.37}$$

where C is the male/female cost ratio. Therefore,

$$R^* = \frac{r_2(n-1)}{r_2 + vnr_1} \xrightarrow[n \to \infty]{} \frac{r_2}{vr_2} = \begin{cases} \dfrac{r_2}{r_1} \text{ when } m = 0 \\ \dfrac{2r_2}{r_1} \text{ when } M = 0. \end{cases} \tag{3.38}$$

Note that the *optimum* ratio of investment we have calculated must be distinguished from the measured value (Equation 3.31). (This is the same as the "equilibrium" value calculated by Benford [1978].) The latter is *defined* as

$$\bar{R} \triangleq \frac{C_M M + C_m m}{C_F F} = \sigma \bar{C},$$

where $\bar{C}$ is the average cost ratio:

$$\bar{C}_2 = \frac{C_M \cdot P + C_m(1-p)}{C_F}.$$

Then, according to Fisher's theory of the equilibrium sex ratio, the reproductive success of females per unit cost should equal that of males:

$$\frac{r_1 S_1}{C_1} = \frac{r_2 S_2}{C_2} \tag{3.39}$$

or

$$\bar{C} = \frac{r_2}{r_1} S \tag{3.40}$$

Multiplying both sides by $\hat{\sigma}$, the equilibrium sex ratio, we have:

$$\bar{R} \triangleq \hat{\sigma}\bar{C} = \frac{r_2}{r_1} S\hat{\sigma}. \tag{3.41}$$

Using the expression derived earlier for the success ratio,

$$S = \frac{\bar{F}}{\bar{M} + \frac{1}{2}\bar{m}} = \frac{2}{\hat{\sigma}(1+\hat{p})},$$

the equilibrium ratio of investment is:

$$\bar{R} = \frac{r_2}{r_1} \frac{2}{1 + \hat{p}} \tag{3.42}$$

This is the same as the *optimum* value R^* when $p = (0,1)$, but has a quite different interpretation. Assuming Fisher's argument holds for a given community, $\bar{R}$ measures what the season-average investment ratio has actually been. R_i^*, on the other hand, is the investment ratio that is optimum from the viewpoint of either the queen, workers, or colony. Thus, $\bar{R}$ can be used as an "index" of which party is "getting its way" in a given situation. The optimal values, R^*, are summarized in Table 3.1.

APPENDIX 3.3

How Can Competitors Coexist Stably?

Consider the following situation. Two opponents, call them 1 and 2, are trying to simultaneously optimize their respective fitnesses, F_1, F_2. Assume that each party can control a strategic parameter u_1 and u_2 so that $F_1 = F_1(u_1, u_2)$ and $F_2 = F_2(u_1, u_2)$. In subsequent chapters the u's will usually be interpreted as proportions of effort allocated to one activity or another. For our purposes here their interpretations are irrelevant. The situation is illustrated graphically in Figure 3.9a where we have sketched constant fitness contours for each party in the "strategy space," (u_1, u_2). The important point to note is that in general each party's fitness depends not only on its own strategy but on the strategy selected by its opponent. Since the two fitnesses do not coincide it is not possible to maximize each simultaneously. Therefore, rather than seek conditions which maximize fitness it makes more sense to look for configurations of stable coexistence. As a candidate, consider the point N where the gradients of the fitness contours are (i) orthogonal, and (ii) parallel to the coordinate axes. More precisely, N is the point (u_1^*, u_2^*) satisfying

$$\frac{\partial F_1}{\partial u_1}\bigg|_{u_2^*} = 0 = \frac{\partial F_2}{\partial u_2}\bigg|_{u_1^*} \tag{3.43}$$

$$\frac{\partial^2 F_1}{\partial u_1^2}\bigg|_{N} \leqslant 0, \qquad \frac{\partial^2 F}{\partial u_2^2}\bigg|_{N} \leqslant 0. \tag{3.44}$$

At such a point the following situation holds. If opponent 1 fixes its strategy at u_1^*, then opponent 2 cannot gain fitness by changing its strategy away from u_2^*. Conversely, opponent 1 finds itself in the same situation: it cannot profit by a unilateral change in strategy away from N. Thus, the point N is stable against "cheating," i.e., small unilateral perturbations. Therefore, N can be considered a competitive equilibrium; in game theory parlance it is called a Nash solution (Owen, 1968). John Maynard Smith has called such strategic pairs (u_1^*, u_2^*) "evolutionarily stable strategies" (ESS) for the following reason. The strategic parameters u_1 and u_2 are reflections of the underlying genetic structure of the opponents, which may be either individuals or populations. Strategic changes arise from genetic alterations (e.g., mutation, recombination) which are generally small perturbations on the existing configuration. At a competitive equilibrium such as N such alterations are at a selective disadvantage and so the point N can be said to be evolutionarily stable.

How Do Cooperators Coexist Stably?

The above definition is predicated on the assumption that simultaneous mutations in both opponents is highly unlikely. However, if there is some mechanism for coordinating strategies it is clear from Figure 3.9a that a simultaneous strategy shift into the shaded region c will benefit both parties. Cooperation will continue to be mutually advantageous until the line P is reached. Along P the fitness gradients are anticolinear (i.e., the fitness contours are tangent):

$$\nabla F_1 = \lambda \nabla F_2, \qquad \lambda \leqslant 0 \text{ along } P \qquad (3.45)$$

P is called the cooperative solution, or Pareto optimum. If we replot Figure 3.9a using F_1 and F_2 as coordinates the cooperative equilibria P corresponds to the fitness frontier $\hat{P}$ shown in Figure 3.9b. $\hat{P}$ shows how the fitnesses component F_1 may be "traded off" against F_2. Levins and others have used fitness

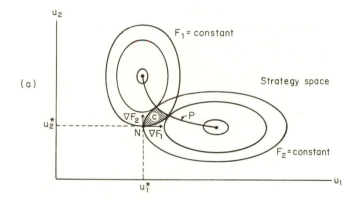

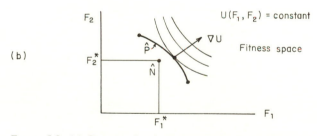

FIGURE 3.9. (a) Contour plots in strategy space of the fitnesses of the two opponents. N is the competitive equilibrium and P is the locus of cooperative equilibria. (b) The cooperative and competitive equilibria in fitness space. A unique equilibrium on P is located by maximizing a utility function $U(F_1, F_2)$; linear and log-linear utilities are shown in the Figure.

space representations to investigate adaptations to heterogeneous environments (Levins, 1968); for our purposes the strategy space representation is more convenient. It is possible to derive the fitness set analytically from basic genetic considerations (Rocklin and Oster, 1976); for our purposes, however, a phenomenological treatment will be sufficient.

Although we shall not concern ourselves with the cooperative solution in this chapter we shall encounter the concept again in Chapter Six. There F_1 and F_2 will be viewed as two conflicting fitness components of the same "player." In order to arrive at a unique equilibrium strategy one must construct a "utility function" $U(F_1, F_2)$ (or "adaptive function" in Levins' 1968 terminology) which reduces the incommensurable fitnesses F_1, F_2 to a common currency. Methods for constructing $U(\cdot, \cdot)$ are generally ad hoc (cf., Chapter Eight); two common models are to combine F_1 and F_2 linearly or log-linearly.

$$U(F_1, F_2) = \alpha F_1 + \beta F_2, \qquad (3.46)$$

$$U(F_1, F_2) = F_1^\alpha F_2^\beta. \qquad (3.47)$$

Then the optimum net fitness is located by solving the programming problem:

$$\text{Max } U(F_1, F_2) \qquad (3.48)$$

subject to F_1, F_2 on $\hat{P}$. $\qquad (3.49)$

There Can Be an Infinite Number of Competitive Equilibria

In the worker-queen conflict case shown in Figure 3.7, conditions of Equations (3.43) and (3.44) defining the competitive equilibrium are not fulfilled anywhere in the feasible set. More generally, the strategic parameters u_1 and u_2 will always be bounded $u_1 < u_1^{\text{MAX}}$, $u_2 < u_2^{\text{MAX}}$. Thus, the feasible region of strategy space will be bounded, and it can easily happen that there is no point in the feasible strategy set which satisfies Equations (3.43) and (3.44). In such cases a more general

114

definition than these must be used. Recall that we defined a stable competitive equilibrium by the condition that unilateral strategy changes cannot increase fitness. That is, (u_1^*, u_2^*) is a competitive equilibrium if:

$$F_1(u_1, u_2^*) \leqslant F_1(u_1^*, u_2^*) \qquad (3.50)$$

and

$$F_2(u_1^*, u_2) \leqslant F_2(u_1^*, u_2^*). \qquad (3.51)$$

Equations (3.43) and (3.44) are sufficient to satisfy these conditions, but they are not necessary. An examination of Figure 3.7 reveals that all of the points on the boundary of the feasible set between the points W and Q satisfy the conditions of Equations (3.50) and (3.51), and so qualify as stable competitive equilibria. The situation is analogous to that of the cooperative (Pareto) equilibria: there are an infinite number of points which are competitively stable, and we require additional information to locate a unique configuration. However, unlike the cooperative case, it is not easy to justify constructing a composite adaptive function since the competing fitness criteria of the opponents are fundamentally incommensurable. There are at least three possible ways the conflict situation may resolve to a unique equilibrium. (1) Other competitive forces may settle the competitive issue between the opponents. In the worker-queen example the queen may be able to employ coercive means such as aggressive behavior or inhibitory pheromones to achieve her fitness optimum. In such cases there may actually be no competition at all if all the power resides in one party's hands. (2) Higher levels of selection may decide the issue. In social insects colony-level selection is probably important and can tip the balance of power in favor of the queen in Figure 3.7, since in that case colony selection is colinear with the queen's interests. (3) Historical accident may decide the outcome. If stochastic events are important, the competitive equilibrium may lie anywhere on the equilibrium set (cf., Chapter Eight).

Competitive Equilibrium When There Are a Finite
Number of Strategies

So far we have considered only the case where each opponent has an infinite number of strategies. That is, the strategic control parameters u_1, u_2 could take on any value on some bounded interval. Frequently, however, one or both opponents have at their disposal only a finite number of strategies. How is the competitive equilibrium defined in this case? The answer is to convert the finite game into a continuous game. We illustrate

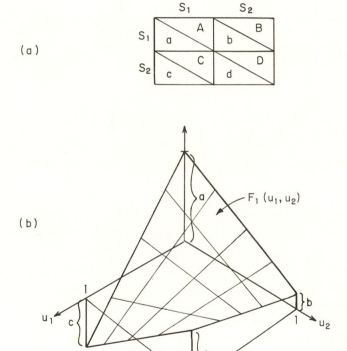

FIGURE 3.10. Competitive equilibria with a finite number of strategies.

this by an example and refer the reader to any text on game theory for a complete discussion (e.g., Owen, 1968; Auslander, Guckenheimer, and Oster, 1978).

Consider the game shown in Figure 3.10a. Each player has but two strategies, which we denote by (s_1,s_2) and (S_1,S_2). The fitness rewards are shown in the "payoff matrix"; rewards a,b,c,d going to opponent 1 and rewards A,B,C,D going to opponent 2. Thus if 1 plays s_1 and 2 plays S_2, then 1 receives a payoff of b and 2 receives a payoff of B. In order to find competitive equilibria we proceed as follows.

Define two continuous strategy parameters u_1 and u_2 which we allow to vary between 0 and 1: $0 \leqslant u_1, u_2 \leqslant 1$. We interpret these parameters as probabilities. $u_1 = 0.7$ means that opponent 1 plays s_1 30 percent of the time and s_2 70 percent of the time; that is $u_1 = 0.7$ is biased 70 percent in favor of strategy s_2. $u_1 = 0$ or 1 corresponds to the "pure" strategies of playing exclusively strategy s_1 or s_2, respectively.

The discrete payoff matrix is converted into a continuous payoff function by linearly interpolating between the payoffs. This is shown in Figure 3.10b where we have graphed the payoff function $F_1(u_1,u_2)$ by considering the matrix entries a,b,c, and d as the height of the surface at the corners of the unit square. A similar surface is constructed for $F_2(u_1,u_2)$. The equation for the payoff functions are obtained from the payoff matrix by constructing the bilinear functions:

$$F_1(u_1,u_2) = (1 - u_1,u_1) \begin{pmatrix} a & b \\ c & d \end{pmatrix} \begin{pmatrix} 1 - u_2 \\ u_2 \end{pmatrix}$$

$$= u_1 u_2 [(a + d) - (c + b)] + u_1 [b - d]$$
$$+ u_2 [c - d] + d,$$

$$F_2(u_1,u_2) = (1 - u_1,u_1) \begin{pmatrix} A & B \\ C & D \end{pmatrix} \begin{pmatrix} 1 - u_2 \\ u_2 \end{pmatrix}$$

$$= u_1 u_2 [(A + D) - (C + B)] + u_1 [B - D]$$
$$+ u_2 [C - D] + D.$$

117

The conditions for an ESS are then obtained as before:

$$u_1^* = (d - c)/[(a + d) - (c + b)], \qquad 0 \leqslant u_1 \leqslant 1$$
$$u_2^* = (D - C)/[(A + D) - (C + B)], \qquad 0 \leqslant u_2 \leqslant 1.$$

Maynard Smith (1976) has noted that discrete strategy games frequently have ESS's which are pure strategies, since it may be difficult to satisfy the above equations for (u_1^*, u_2^*) in the unit square.

CHAPTER FOUR

The Biological Characteristics of Caste Systems

We have defined a caste intuitively as a set of individuals, smaller than the society itself, that is specialized to perform one or more roles. Because this is a purely functional definition based on the behavior of sets of individuals, it is difficult to express in quantitative terms. If we can find other, more easily measurable characteristics that correlate well with the behavioral roles—that is, if we can use "markers"—then the task of empirically determining caste characteristics will be greatly facilitated.

In fact, this is easily accomplished in the social insects. The most fundamental cleavage within the colony is between the reproductive caste and the sterile worker caste. Within the worker caste one can further classify individuals into age and size classes. It turns out that this classification does correlate well with the various tasks each individual takes upon herself to perform. Thus, it is useful to speak of "temporal castes" whose behavior is age-correlated and "physical castes" whose behavior is size-correlated. Temporal castes are most strongly developed in the ants, stingless bees, and honeybees. In the halictine bees, bumblebees, wasps, and termites the correlation between age and behavioral role is less pronounced. In most termite species and many ant species the size, or more precisely the allometric proportion of individual workers, is a good predictor of their behavioral repertory. Later we shall introduce a third caste marker, "tempo," defined as the activity level by which various caste behavior patterns can be characterized.

Together, age, size, and tempo provide an adequate set of caste descriptors to commence the study of ergonomic optimization.

4.1. BEHAVIORAL ACTS, ROLES, AND CASTES

The fundamental distinctions among individual workers in a colony are based on the actual tasks they perform and the various roles they play in accomplishing the functions of the colony as a whole. An example of the behavioral repertories of two physical castes is given in Table 4.1.

Lists of distinct behavioral acts are made from protracted observations of individuals, selected sets of individuals, or the entire colony. If distinct physical and temporal groups can be distinguished, separate repertories for each are compiled. In the unlikely event that some of these arbitrarily defined categories prove to have identical repertories, they cannot be distinguished as separate castes—and the data concerning their repertories can be subsequently combined. The next step is to compute the relative frequencies of the acts and to fit them to standard distributions such as the lognormal Poisson or negative binomial. This method has been used with success by Fagen and Goldman (1977) to make estimates of total repertory size even before a complete catalog is actually compiled. An observer thus can obtain a measure of the adequacy of his sample size. For example, after 1,222 separate behavioral observations had been made on the minor worker caste of a colony of *Pheidole dentata*, 26 kinds of behavioral acts could be recognized; the latter constituted the known behavioral repertory of this caste. By fitting the frequency data (see Table 4.1) to a lognormal Poisson distribution, the actual number of kinds of behavioral acts was estimated to be 27, with a 95 percent confidence interval of (26, 28). The major workers were observed performing 8 kinds of behavioral acts; the true number was estimated to be 9, with a confidence interval of (8, 10).

TABLE 4.1. A behavioral repertory: behavioral acts by the two physical castes of the ant *Pheidole dentata* in an undisturbed colony are listed and their relative frequencies given. N, total number of behavioral acts recorded in each column. (From Wilson, 1976a.)

	Frequency of Behavioral Acts	
Behavioral Act	Minor workers ($N = 1222$)	Major workers ($N = 208$)
Self-grooming	0.18003	0.56373
Allogroom adult:		
Minor worker	0.04992	0
Major worker	0.00573	0
Alate or mother queen	0.01146	0
Brood care:		
Carry or roll egg	0.01391	0
Lick egg	0.00245	0
Carry or roll larva	0.12357	0
Lick larva	0.09984	0.02941
Assist larval ecdysis	0.00409	0
Feed larva solid food	0.00573	0
Carry or roll pupa	0.03601	0
Lick pupa	0.01882	0
Assist eclosion of adult	0.00818	0
Regurgitate:		
With larva	0.02128	0
With minor worker	0.03764	0.22059
With major worker	0.00573	0
With alate or mother queen	0.00327	0
Forage	0.12111	0.02941
Feed outside nest	0.04337	0.01471
Carry food particles inside nest	0.05237	0
Feed inside nest	0.05810	0.01471
Lick meconium	0.00573	0
Carry dead nestmate	0.01882	0.04902
Carry or drag live nestmate	0.00246	0
Eat dead nestmate	0.06383	0.07843
Handle nest material	0.00655	0
Totals	1.0	1.0

The next step is the construction of an *ethogram*, which incorporates not only the repertory of a caste but also the transition probabilities connecting individual acts and the time distributions spent in each act. When the ethogram also takes into account the interactions of parents and offspring, dominant and subordinate males, and other members of the society, the description can be conveniently referred to as a *sociogram*. Ethograms can cover all of the repertory or certain well-defined portions of it. Such quantitative studies have been conducted on ants, honeybees, hermit crabs, mantis shrimps, and rhesus monkeys, with promising results (see reviews by Dingle, 1972; Wilson, 1975a:194–200).

Figure 4.1 illustrates an imaginary example of an ethogram with roles and castes delimited. For any individual, certain of the behavioral acts (a_i) will be linked together by relatively high transition probabilities: in *Pheidole dentata*, for example, pupal licking is associated with high frequency to pupa carrying, egg licking, and egg transport; nest construction gives way with high probability to foraging outside the nest; and so on. A set of closely linked behavioral acts can be defined as a *role*, even if the acts are otherwise quite different. It is generally true, for example, that the act of grooming the queen is closely linked with the very different acts of regurgitating to the queen and removing freshly oviposited eggs. All of these responses can be considered part of the single role of queen care. Finally, we can define a *caste* as a group that specializes to some extent on one or more roles. Frequently the term *task* is used to denote a particular sequence of acts that accomplishes a specific purpose, such as foraging or nest repair. Ordinarily a task is identical to a role or is composed of the subset of acts within a role, but it might conceivably consist of acts distributed across two or more roles.

The information in the ethogram can be organized in a mathematically useful structure as follows. Denote the distribution of times spent in each state (act) by $w_i(\tau)$. That is, if T_i is a

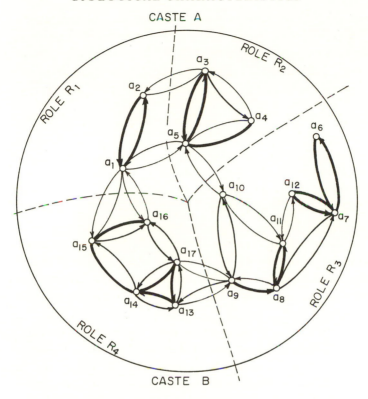

FIGURE 4.1. A form of ethogram by which roles and castes can be more precisely defined. Each set of behavioral acts a_i linked by relatively high transition probabilities constitutes a role. The thicknesses of the connecting lines indicate the magnitudes of the transition probabilities. When a distinct group (for example, a size or age cohort) attends preferentially to one or more roles, it is defined as a caste. Roles can also consist of overlapping sets of behavioral acts; for convenience of graphical representation they are shown here as disjoint sets.

random variable giving the amount of time an individual spends performing act a_i, then $\mathrm{Prob}[\,T_i = \tau\,]$ has a density function $w_i(\tau)$. The transition probabilities between a_i and a_j we denote by $p_{ij}(\tau)$ to indicate that the likelihood of switching to a_j may be conditional on the amount of time already spent in a_i. Then the matrix of holding times $\mathbf{W}(\tau)$ together with the matrix of conditional transition times, $\mathbf{P}(\tau)$, constitute a semi-Markov process (Howard, 1971). In Chapters Six and Seven we will exploit this representation of the ethogram to calculate caste "efficiencies," a measure central to our ergonomic optimization models.

4.2. THE NATURE OF PHYSICAL CASTES

The functional definition of caste just given is clearly the most relevant with respect to the evolution and ecology of social species. However, ethograms are tedious and difficult to construct, making it desirable to seek more operational quantities. The most striking distinction between individuals in many species is their physical appearance. Furthermore, empirical studies have shown that there is apparently always a close association between allometric proportions and behavioral roles. In this section "caste" will be used as an abbreviation of "physical caste."

During the past twenty-five years a large amount of experimental work has been conducted on the determination of caste in individual social insects. In nearly every species thus far investigated, caste determination has proved to be environmental rather than genetic in nature. In most instances either the amount and quality of food given the immature forms or the pheromones to which they are exposed, or both, are among the decisive influences. In other words, each individual starts its development fixed genetically only to sex; it is otherwise totipotent with reference to caste (general reviews are presented by Wilson, 1971; and Schmidt et al., 1974). Only two exceptions are known, and these are strongly qualified. The first is the stingless bee genus *Melipona*, in which queens appear to be the full heterozygotes of an independently segregating two-locus system, and hence regularly to constitute a randomly determined one-fourth of the female population (Kerr, 1950). In this case, however, significant deviations can be obtained in either direction by strong under- or overfeeding (Kerr and Nielsen, 1966; Kerr, 1974; Darchen and Delage-Darchen, 1974). Thus, genetically determined queens tend to be converted into workers whenever food is scarce, a strategy of obvious benefit to the colony.

The second exception is the European slave-making ant *Harpagoxenus sublaevis*, whose ergatomorphic (workerlike) reproductive females differ from typical winged queens by what appears to be a single recessive allele (Buschinger, 1975). However, the difference is one that usually occurs between colonies and is not ordinarily a basis for caste differentiation within colonies. Furthermore, winged queens, which are the homozygous recessives, have been found in only 3 colonies out of more than 500 examined in Germany by Buschinger.

Termites have an almost fully discretized caste system. Because the development of these insects is hemimetabolous (gradual, and not entailing major differences between the life stages), the immature colony members are basically similar in body form to the adults and capable of performing many of the same behaviors. Consequently not only the adults but also young insects belonging to some of the juvenile instars can function as discrete castes. In many of the species of "higher" termites, which comprise the phylogenetically advanced family Termitidae, other caste differences are based on sex. One of the more complex examples is shown in Figure 4.2. It can be seen that hemimetabolous development has been placed in the service of caste differentiation. For example, a male can become a small worker and then proceed on through two molts to become a large soldier. (In order to become a small soldier it must follow another developmental track.) Although documentation is lacking in this particular species, each of the remaining ten instars has at least the potential to perform differently within the colony and hence by definition comprise a separate caste.

The social Hymenoptera (wasps, ants, and bees) are holometabolous in development, meaning that the larva is a grublike

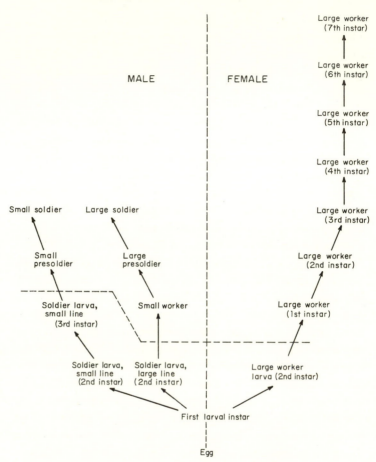

FIGURE 4.2. The discretization of castes is enhanced in termites by the hemimetabolous development of these insects. The diagram above shows the full developmental pathways of a species of *Trinervitermes*, one of the phylogenetically advanced members of the Termitidae with a relatively complex caste system. Each time an individual molts it enters a new and physically larger instar. Females can eventually develop into large workers, and males into small workers, small soldiers, or large soldiers. Termites still in the larval instars are helpless, but those in each of the later instars are capable of more independent action and have at least the potential of serving as a distinct caste. (Based on Noirot, 1969.)

form radically different from the adult. Transformation to the adult form is accomplished by passage through a quiescent pupal stage. In a few hymenopterous species the larvae function as a distinct caste. Those of the genera *Vespa* and *Vespula* supply the adult workers with salivary secretions rich in trehalose, glucose, and other carbohydrates, which they metabolically convert from the tissues of the prey insects fed to them (Maschwitz, 1966; Ishay and Ikan, 1969). Division of labor is especially advanced in the Oriental hornet (*V. orientalis*): the adults are incapable of gluconeogenesis and depend on the larvae for the sugars vital to their daily energy needs (Ishay and Ikan, 1969). Workers of pharaoh's ant (*Monomorium pharaonis*) collect salivary secretions from the larvae, and during times of extreme dryness survive longer if given access to a small number of these immature forms (Wüst, 1973). Queens of the ant *Leptothorax curvispinosus* spend much of their time "grazing" from the secretions of one larva after another, and it appears probable that they receive much of their sustenance in this way (Wilson, 1974b). Went et al. (1972) have suggested that larvae of *Manica* and *Veromessor* are responsible for ingesting solid food and sharing it with the adults, but the evidence they present is indirect and less than conclusive. Larvae of weaver ants in the genus *Oecophylla* produce the silk used in nest construction (Hemmingsen, 1973).

Clearly, then, the larval stage, embracing all of the active immature instars, can constitute a distinct physical caste in social Hymenoptera. But this is not a universal phenomenon. Stingless bee larvae are sealed in brood cells throughout growth, while larvae of the harvesting ant *Pogonomyrmex badius* never exchange liquids with adults in either direction. Furthermore, when hymenopterous larvae contribute labor, it appears to be primarily in the domain of specialized metabolism and food storage and exchange. Perhaps this last restriction is to be expected, given the near immobility of the immature stages. Nevertheless, the subject of "child labor" in the social Hymenoptera has scarcely been explored, and many surprises

127

may await us. It is at least suggestive that striking differences in shape, pilosity, and glandular development exist from one larval instar to another in a few ant genera, including *Pachysima* (Wheeler, 1918), *Crematogaster* (*Nematocrema*) (Delage-Darchen, 1972), and *Pheidole* (Passera, 1974). The possibility of an associated division of labor is worth examination.

In spite of these intriguing facts and clues, there can be no doubt that by far the greatest caste differentiation in the Hymenoptera is found among the adult instars. The wasps, ants, and bees have all made use of temporal polyethism, that is, changes in behavior as a function of age during the instar. This will be discussed at length below. In the ants there has been in addition an anatomical differentiation of adults. Physical castes in ants are roughly comparable in diversity to those of the termites, but their physiological provenance is very different. In the first place, there is no sexual differentiation as in the higher termites. Only female ants serve as truly functional workers. Although males of at least two species of carpenter ants, *Camponotus herculeanus* and *C. ligniperda*, exchange liquid food with other members of the colony (Hölldobler, 1966), members of this sex are, by and large, passive drones that remain in the colony only for a short while before departing on the nuptial flight. Also, anatomical diversification among the females is based on allometric growth. During larval development, the imaginal discs (patches of undifferentiated tissue destined to be transformed into adult organs at the pupal state) grow at different rates, a process that swiftly accelerates during pupal development (Brian, 1957b, 1965b; Schneirla et al., 1968; Lüscher et al., 1977).

The principal result of differential growth rate in the imaginal discs is that various organs end up with different sizes relative to one another according to how large the individual is at the termination of the larval period. That is, the final adult size determines how much overall growth the various organs have attained. Thus, if the disc destined to transform into part of

128

the head is growing faster than the disc destined to transform into part of the thorax, it will finish proportionately larger in an ant that attains a larger overall size. In short, big ants will have proportionally even bigger heads. If each disc grows exponentially, and if the disc growth rates do not change much in the course of development, the sizes of two parts will be related by a simple power law: $\log y = \log b + a \log x$, or, equivalently, $y = bx^a$ where y and x are linear measures of the two body parts and a and b are fitted constants the values of which depend on the nature of the measurement taken. This simple relation is referred to as *allometry* or heterogonic growth (Huxley, 1932; Gould, 1966). On a double logarithmic plot the curve is a straight line. Its slope a is determined by the rate of divergence of the two body parts with increasing total size and can be referred to as the allometric constant. If a is equal to unity, no divergence takes place with an overall increase in size; the growth is then referred to as *isometric*. The greater the departure of a from unity, the more striking the differential growth. Skellam, Brian, and Proctor (1959) found that in *Myrmica rubra* the imaginal discs of the wings and legs actually grow in this elementary fashion during adult development of queens and males. The organs predictably conform to the basic allometry equation in their final adult form.

It is possible to produce a wide array of deviations from elementary allometric growth by simply speeding or slowing the growth rates of different discs according to different time schedules. Further complexity can be introduced by making the rates dependent on the total size of the larva reached by certain ages. This last effect is crucial in the determination of the queen and worker castes in *Myrmica*, as documented by M. V. Brian. It also occurs during the differentiation of the worker caste of the *Eciton* army ants, as discovered by Tafuri (1955). These modifications are crucial for the discretization of physical castes within the adult instar of ants, a process that will now be explored in some detail.

129

4.3. THE EVOLUTION OF PHYSICAL CASTES IN ANTS

Wilson (1953) demonstrated that the allometry equation, or relatively simple modifications of it, can be applied amost universally to continuous variation in the hard parts of ants. The comparative study of allometry has proved fruitful in tracing the evolution of castes. Polymorphism, as this research has led us to understand it, embraces two variable characters in the adult females of any species: the allometric growth series and the size variation among the adult members of individual colonies. A physical caste system (or polymorphism as it is also frequently called) is defined as nonisometric relative growth occurring over a sufficient range of adult size variation to produce individuals of distinctly different proportions. Where polymorphism exists, it has always been found to be closely linked to division of labor.

By comparing a large fraction of the more than 10,000 living ant species, it has been possible to infer five major steps in the evolution of physical castes (Wilson, 1953):

(1) *Monomorphism*

The workers of the normal mature colony are either isometric or else display very limited size variability, or both. A plot of their size-frequency distribution is symmetrical and has only a single mode. In other words, the properties of variation are not basically different from those in a typical random collection of nonsocial insects. The worker castes of most ant genera and species are monomorphic. Also, within the majority of genera and higher taxonomic groups monomorphism is evidently the primitive state.

(2) *Monophasic Allometry*

The relative growth is nonisometric, meaning that the allometric constant *a* is greater or less than unity. In the most elementary form of monophasic allometry, and hence of worker polymorphism generally, feeble nonisometric variation is displayed over a short span of size variation; this variation in turn is grouped around a single mode with possible skewing in the direction of the major caste. A more advanced stage involves an increased variation in size together with a marked tendency toward bimodality; this is exemplified by the carpenter ant *Camponotus castaneus* depicted in Figure 4.3.

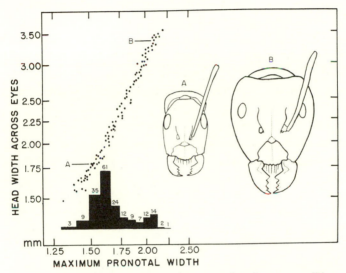

FIGURE 4.3. Physical caste systems in ants are based on two qualities, nonisometric growth and increased size variation among the adult females of each colony accompanied by a tendency toward bi- or trimodality. The workers of the carpenter ant *Camponotus castaneus* depicted here possess an elementary form of polymorphism. Some of the body parts are nonisometric; in other words, they increase or decrease in relative size as total body size is enlarged. Head width increases faster than pronotal width, while pronotal width is isometric with reference to most of the rest of the body. (The pronotum is the upper plate of the first segment of the thorax just visible behind the head in the drawing.) The allometry of *C. castaneus* is "monophasic," meaning that the slope of the relative growth curve remains constant or nearly so. The size-frequency curve is weakly bimodal. The small-headed individuals clustered around the smaller mode are called minor workers, those with large heads around the larger mode are the major workers, and those around the midpoint between the modes are the media workers. (From Wilson, 1953.)

(3) Diphasic Allometry

The allometric regression line, when plotted on a double logarithmic scale, "breaks" and consists of two segments of different slopes that meet at an intermediate point. In the several species known to possess this condition, including the leafcutter ant (*Atta*) and African driver ants (*Dorylus*), the size-frequency curve is bimodal, and the saddle between the two frequency modes falls just above the bend in the allometry curve. Diphasic allometry permits the stabilization of the body form in the small caste while providing for the production of a markedly divergent major caste by means of a relatively small increase in size. The lower segment of the relative growth curve is nearly isometric, so that individuals falling within a large segment of the size range are nearly uniform in structure; but the upper segment leading to the major caste is strongly nonisometric, with the result that a modest increase in size yields a new morphological type.

(4) Triphasic and Tetraphasic Allometry

The allometric line breaks at two points and consists of three straight segments. The two terminal segments, representing the minor and major castes respectively, deviate only slightly from isometry while the middle segment, encompassing the media caste, has a very steep slope. The effect of triphasic allometry is the stabilization of body proportions in the minor and major castes. An example is presented in Figure 4.4. Recently Baroni Urbani (1976) has reported a case of tetraphasic allometry in the antennal length of the West African ant *Camponotus maculatus*. The curve resembles that of triphasic species, except that in the largest size classes a high degree of allometry is resumed.

(5) Complete Dimorphism

Two morphologically very distinct size groups exist, separated by a gap in which no intermediates occur. Each class is

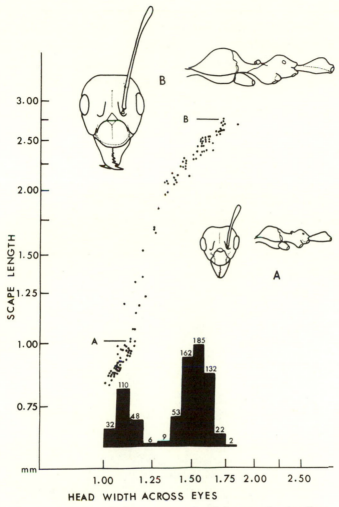

FIGURE 4.4. In the Asiatic weaver ant *Oecophylla smaragdina* the allometry is triphasic. Three different slopes are possessed by such allometric characters as the length of the scape (first antennal segment) taken as a function of body size, in this case represented by the maximum width of the head. The size-frequency curve is strongly bimodal, with majors predominating. The heads and middle body parts of selected minor and major workers are also depicted. (From Wilson, 1953.)

nearly isometric, but the allometric regression curves are not aligned, a condition suggesting that complete dimorphism can arise directly from triphasic allometry. Examples include most queen-worker differences in ants and many minor-major divisions in a wide phylogenetic scattering of genera.

By arranging species of ants along a gradient from what appear to be the simplest to the most advanced systems, worker castes can be found that display virtually every conceivable step in a transition from perfect monomorphism to complete dimorphism (see Figure 4.5). Certain large taxonomic groups, such as the subfamilies Myrmicinae and Formicinae, embrace the entire evolutionary sequence within themselves. The evolution has thus occurred repeatedly within multiple phyletic lines in the ants and produced a remarkable degree of convergence between the lines.

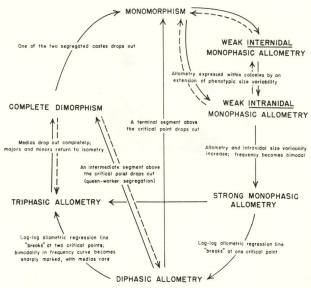

FIGURE 4.5. Inferred pathways in the evolution of physical caste systems in ants. (From Wilson, 1953.)

Of the two principal qualities of polymorphism, the size-frequency distribution has been subject to the stricter and more notable convergence. When individual colonies of a given species show a slight increase in size variance, the frequency curve is almost always skewed toward the larger size classes. When the intracolonial size variance is still greater, the frequency curve is bimodal. In at least one species of army ant with extreme size variation, data published by Topoff (1971) and da Silva (1972) show the existence of three size modes. But otherwise enlarged intracolonial size variation is typically associated with bimodality, with the large workers being less common. The caste system of *Camponotus castaneus*, illustrated in Figure 4.3 above, is typical of this condition.

The ways by which ant species create castes out of the adult instar are few. It is reasonable to ask how a colony limited to simple skewing and a maximum of two or three modes is able to produce castes in proportions that match the frequencies of numerous environmental contingencies. We believe the answer to be that the evolutionary sequence unfolds within narrow physiological constraints. Species can only improve their situation to the extent of increasing the intracolonial size range and arranging something close to the optimum numbers of majors, minors, and medias. (This optimization problem will be discussed in Chapter Six.) Physical caste systems, in a word, are coarse-tuned rather than fine-tuned.

4.4. DEVELOPMENTAL SHAPING OF THE CASTE SIZE-DISTRIBUTION

We will now suggest how *growth transformation* can account for the observed size-frequency distributions, and show how this process can be adjusted in simple ways to generate a substantial array of caste systems (see also Alberch et al., 1978).

Consider first the production of a small class of majors by skewing of the adult worker size-frequency distribution. As

135

suggested in the schema of Figure 4.6, newly hatched larvae vary in body weight, probably according to a normal distribution. It is likely that they also vary in other qualities, such as the amount of yolk available to them during embryonic development, the temperature at which they developed, and so forth. Suppose that some of these factors influence the final body size

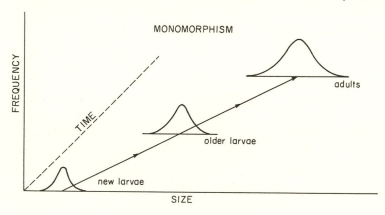

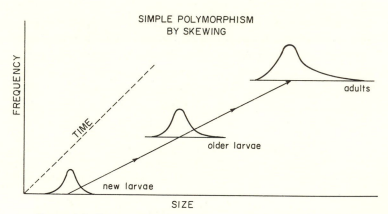

FIGURE 4.6. Simple caste systems are created during the evolution of ant species by the skewing of the size-frequency curve toward the larger size classes, accompanied by allometric variation. This can be achieved by a growth transformation function that converts small differences in initial larval size, or any other factor affecting caste, into disproportionately greater differences in the final larval size.

attained by the larvae and hence by the adult ants. If this growth transformation remained constant with increases in initial size, the result would be a set of workers whose size-frequency distribution is approximately normal. Such distributions are typically accompanied by isometry, and consequently the entire worker caste will be monomorphic. Suppose, however, that the transformation is such that the larger the initial size of the larvae (or the greater the quantity of other caste-biasing factors present), the more rapid its growth, so that individuals starting large finish proportionately larger. Such a relationship might be described as follows: $y_t = F[y_0]$ where y_0 and y_t are the body size (or quantity of other caste-biasing factors) at the start and finish of larval growth. If the transformation is linear, the result would be monomorphism, in conformity with the conditions just described. Where the transformation is exponential, so that the subsequent rate of growth increases as a function of the starting point y_0, the result will be a size-frequency distribution skewed toward the upper size classes. The magnitude of the exponential constant represents the sensitivity of larval growth to the initial conditions encountered by the young larva. Alternatively, the constant can depend on conditions encountered later in larval life, with similar final results. By relatively small adjustments in the transformation function, the caste system of a species can be conspicuously altered.

The next simplest conceivable step in the elaboration of size-frequency distributions is the introduction of decision points during development, as illustrated in Figures 4.7 and 4.8. The decision point is a time in development at which one or the other of two sets of growth constants are acquired by the immature form. Thereafter the individual will proceed in its development toward one subcaste or another, with little chance of deviation. Decision points are efficient devices for sorting colony members into two or more independent populations. The rules for shunting individuals into one direction as opposed

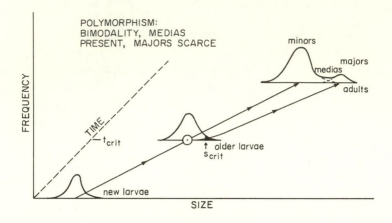

POLYMORPHISM:
BIMODALITY, MEDIAS
PRESENT, MAJORS SCARCE

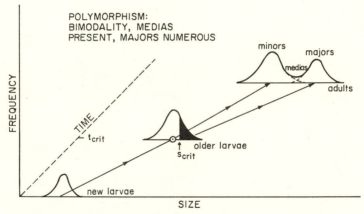

POLYMORPHISM:
BIMODALITY, MEDIAS
PRESENT, MAJORS NUMEROUS

FIGURE 4.7. When a switching point is introduced at a critical developmental time (t_{crit}), larvae that have attained a threshold size (s_{crit}) increase their growth rate and move toward the major worker mode of the final adult size-frequency distribution. (The fraction of larvae destined to travel this divergent pathway are indicated by the shaded portion of the middle frequency curves.) By adjusting s_{crit}, species can set the percentage of major workers, as illustrated by a comparison of the upper and lower diagrams. In both of these imaginary examples, the divergence of the two developing pathways is slight enough so that the final minor and major frequency distributions overlap, producing a media class.

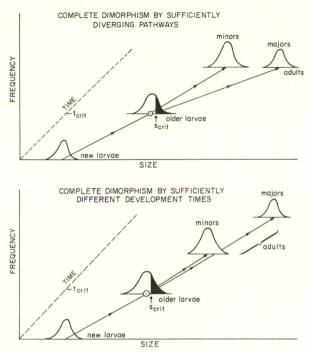

FIGURE 4.8. This schema is an extension of that presented in Figure 4.7. When the two developmental pathways diverge rapidly enough (*above*) or the larger larvae continue their development for a sufficiently longer period of time in comparison with the smaller larvae (*below*), the final size-frequency distributions are disjunct and the worker caste is completely dimorphic.

to the other can be adjusted so as to regulate the relative sizes of the two populations. The examples given in Figure 4.7 utilize threshold size at a critical developmental time as the shunting rule; this is the situation in queen-worker determination in the ant *Myrmica ruginodis* (Brian, 1955) and the bumblebee *Bombus rufocinctus* (Plowright and Jay, 1978). Larvae that attain a certain size by a critical time continue rapid development toward a still larger ultimate size; those that fall short proceed at a distinctly slower rate and are even more behind at

139

the completion of adult development. By setting the threshold size low or high relative to the usual size-frequency distribution of larvae at the critical development time, species can arrange a lower or higher proportion of the larger caste.

Shunting rules are based on token stimuli, which species have appropriated to adjust the caste ratios. No less than six classes of such stimuli have been identified in various ant species as influencing the determination of individual females to the worker caste as opposed to the queen caste. Two of these are also known to be potent in minor-major worker determination; the latter process is still relatively unexplored and will probably be shown to be subject to other influences as well (see reviews in Brian, 1965b; Wilson, 1971; and Schmidt et al., 1974).

(1) *Larval Nutrition*

Competition among larvae alone might produce the bimodal size-frequency curves that underlie most queen-worker and minor-major distinctions in ant species. Larvae that attain a threshold size by a critical developmental time are shunted toward the larger caste.

(2) *Winter Chilling*

Intraovarian eggs of *Formica* and larvae of *Myrmica* that have been chilled have a greater tendency to develop into queens, an apparent device for timing the emergence of queens in the spring or early summer. Other responses to temperature, humidity, or photoperiod could produce crops of queens (or even worker subcastes) at other times. These would depend on local climatic conditions and the idiosyncratic features of the colony life cycle in each species.

(3) *Temperature*

The larvae of *Formica* and *Myrmica* tend to develop into queens more readily if reared at higher temperatures.

140

(4) Caste Self-Inhibition

The presence of a mother queen inhibits production of new queens in *Myrmica*, *Monomorium*, and *Oecophylla*; likewise the presence of soldiers inhibits the production of soldiers in *Pheidole morrisi*. This negative feedback loop could obviously serve to stabilize caste ratios when there is a need to fix the ratio instead of making it flexible in response to short-term needs.

(5) Egg Size

In *Formica* and *Myrmica*, the larger the egg, the more yolk and the more likely the larva is to develop into a queen as opposed to a worker. No information is available on the relation of egg size to minor-major determination.

(6) Age of Queen

Young queens of *Myrmica* tend to produce more workers; the queen's age could, of course, be reflected in the size of the eggs she lays. Smaller egg size in the batch produced by a nest-founding queen might further explain the occurrence of nanitic workers in the first brood; in other words, egg size could evolve to insure the production of nanitics at this stage in the colony life cycle.

Almost all bees, wasps, and termites also employ subsets of the basic six factors just listed, insofar as our limited knowledge allows us to judge. Some factors are merely biasing in their effects, rendering an individual more likely to take one direction as opposed to the other upon reaching the point of bifurcation. Others exert their influence directly at the decision points themselves. Often one factor is epistatic with reference to another in the following manner: if condition *a* prevails earlier instead of *a'*, then the larva can respond to either *b* or *b'*. For example, mature larvae of *Myrmica ruginodis* subjected to winter chilling have the capacity to develop into either queens or workers. But only those that subsequently reach a weight of

141

3.5 mg within approximately eight days after the start of posthibernation development actually become queens. Larvae not exposed to chilling always develop into workers, regardless of their size.

The scheduling of receptiveness to caste-biasing stimuli almost certainly represents an idiosyncratic genetic adaptation on the part of each individual species. The later the decision point, the more flexible is the system, in the sense that it permits the colony to make rapid adjustments in the caste ratios. This would seem *a priori* to be of special advantage to species that possess a soldier caste subject to occasional heavy mortality. The older the larvae are when shunted to the soldier developmental pathway, the shorter the time required to fill gaps created by casualties. It is consequently of interest that in *Pheidole pallidula* the point of soldier-*vs*-minor worker determination is in the third and final instar (Passera, 1974). R. A. Metcalf (personal communication) has evidence of a similar timing in *P. dentata*. Third instar larvae of this species also appear to undergo an interval of arrested development prior to reaching the decision point, a phenomenon which, in effect, shortens the response time of the colony as a whole. The timing of queen-worker bifurcation varies greatly among species. In *Myrmica ruginodis* it is very late—about a week prior to the cessation of larval growth (Brian, 1955). But in *Formica polyctena* and the honeybee *Apis mellifera* the opposite is true: a larva is determined to queen or worker within 72 hours after hatching from the egg (Bier, 1958; Schmidt, 1974; Weaver, 1957; Rembold, 1974). A similarly wide variation in timing has been observed among species of bumblebees (Röseler, 1974; R. C. Plowright, personal communication). The relation of these differences to the ecology of the species awaits investigation.

The mode of regulation of the ratios of minor and major workers is also poorly understood. When Passera (1974) altered the percentages of *Pheidole pallidula* major workers (soldiers) from the usual 3–5 percent, most returned to this level within

142

75 days through the differential production of minors and majors. Even colonies containing 50 or 100 percent majors at the outset dropped to an average level of 20–25 percent and were still declining at the end of the experiment. A similar inhibition has been demonstrated in *Pheidole morrisi* by Gregg (1942). These experiments show that minor/major ratios cannot be set entirely by an automatic shunting of larvae at the critical developmental time, as suggested in the model of Figure 4.7. Feedback controls have been added that can temporarily decrease or increase the proportion shunted. The controls might conceivably act by increasing the growth rate of larvae prior to the decision point, by changing the threshold size, or both. Such alterations could be mediated entirely through changes in the endocrine systems of the growing larvae themselves. The endocrine changes are likely to be initiated in turn by the receipt of appropriate token stimuli, such as the increase or decrease of inhibitory pheromones from the major caste or an alteration in the kinds of food given the larvae by the workers. Similar processes have already been documented in the determination of the queen and worker castes of ants and termites, and there is some evidence to suggest their influence in soldier determination in the lower termites (Hrdý, 1972; Lüscher, 1974).

4.5. THE NATURE OF TEMPORAL CASTES

The adult workers of almost all kinds of social insects change roles as they grow older, ordinarily progressing from nurse to forager. Each species has its own distinctive pattern of temporal polyethism, and in many the behavioral changes are accompanied by patterned shifts in the activity of exocrine glands. For example, as honeybee workers shift during a two-week period from a preoccupation with brood care and nest work to an emphasis on foraging, the activity of the hypopharyngeal and wax glands declines somewhat while that of the labial glands increases. For the purposes of ergonomic analysis it is useful to consider different age groups as constituting distinct "temporal" castes (Wilson, 1968). Just as a species may manipulate its own developmental biology in the course of evolution to produce optimum ratios of physical castes, it can adjust the program of role change during adult life to approach optimum ratios of temporal castes.

Two extreme alternatives are open to a species in the process of evolving temporal castes. These are represented in Figure 4.9. The aging period depicted is that which occurs from the moment of the worker's eclosion from the pupal skin to the moment of its death by senescence. The division of the life span into six periods in this imaginary example is arbitrary. The worker undergoes physiological change with age such that its responsiveness to various environmental stimuli changes. For example, suppose that T_1 is the responsiveness to a misplaced egg: the curve indicates that when the worker is very young (age I) it is likely both to be in the vicinity of the egg and to react by picking the egg up and putting it on an egg pile. Its location and/or behavioral responsiveness change as it ages in such a way that its probability of response to the contingency drops off rapidly after ages I or II. Such age-dependent responsiveness has been amply documented in ants (Topoff et al., 1972; Cammaerts-Tricot, 1974; Jaisson, 1975).

Let us now consider the possibilities. In the upper diagram of Figure 4.9, labeled Model 1, the response curves to four contingencies (T_1 through T_4) are all out of phase. The curve of response to T_1 (misplaced egg) is different from the curve of response to T_2 (say, a hungry larva), and so on. As a result, the ensemble of age groups, represented on the right-hand side by the frequency distribution of workers in different age groups that attend to task T_1, is different from that attending to T_2, and so on. As the number of contingencies is increased, and their response curves are all made discordant, there will be one age-group ensemble for each task. Let us now define an age-group ensemble as a *temporal caste*. In the extreme case represented in Model 1 there is a caste for each task. However, the

144

MODEL 1

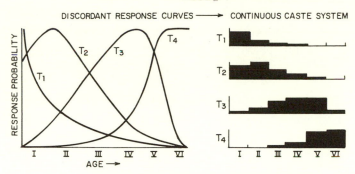

MODEL 2

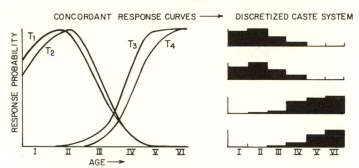

FIGURE 4.9. These two schemas represent the extreme alternatives open to an ant species in the evolution of temporal castes. The age of the adult worker (or of any instar that can function as a worker) is arbitrarily divided into six periods. In the first model (*upper*) the responsiveness of the worker to each of four tasks (T_1 through T_4) changes markedly out of phase with reference to its responsiveness to the other contingencies. As a consequence each of the four tasks is addressed by a distinct ensemble of age groups (temporal castes) which are represented on the right by the frequency distributions of workers in different age groups attending to the contingency. If the number of contingencies (tasks) were increased substantially, the overlap of the age-group ensembles would increase to a corresponding degree and the resulting system would approach a continuous transition. In the second model (*lower*) the response curves are clustered into groups that are approximately in phase, resulting in more than one contingency being addressed by the same age-group ensemble (caste). Two ensembles perform two tasks each. Even if the number of contingencies were increased substantially, the number of ensembles would remain small. (From Wilson, 1976a.)

distinction between age-group ensembles will be blurred as more tasks are added. The overlap in the age-group frequency curves is so extensive that after ten or so contingencies are added, the system becomes effectively continuous. For this reason we suggest that such an arrangement be called a *continuous caste system*.

The approach to continuity in Model 1 is marked by complexity and subtlety. The evolving ant species can easily adjust the programming of individual worker responsiveness to attain discordance, which in turn yields one caste specialized for each task. Thus, only a relatively elementary alteration in physiology is needed to produce a complex caste system.

Next consider the alternative option, depicted in the lower half of Figure 4.9. Here various of the response curves are concordant, or at least approximately so. As a result the same statistical ensemble of workers attends to more than one task. As the number of tasks increases, the number of castes does not keep pace; conceivably it could remain low, say corresponding to as few as 2 or 3 distinct ensembles. Thus, the species has chosen to operate with a *discrete caste system*—comprised of a relatively few, easily recognized age-group ensembles. The evolutionary process leading to such a system can be called "behavioral discretization" (Wilson, 1976a). It can operate through physiological alterations as potentially simple as those that yield continuous caste systems.

Few studies of temporal polyethism have been designed in a way that permits a consideration of the hypothesis of discretization. A recent analysis of the ant *Pheidole dentata* by Wilson (1976a) showed that the system has been sharply discretized. Virtually all of the 26 behavioral acts recorded in the minor worker caste can be divided according to three age periods in which they are performed approximately in concert. Fifteen of the acts are represented in Figure 4.10. Data presented by Higashi (1974) on the Japanese wood ant *Formica yessensis* indicate that temporal castes have also been discretized

146

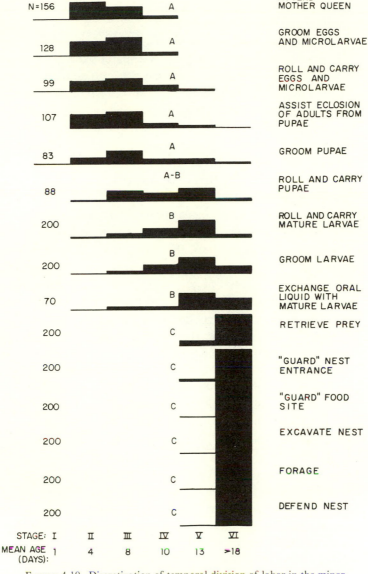

FIGURE 4.10. Discretization of temporal division of labor in the minor workers of the ant *Pheidole dentata*. The proportions of workers of six arbitrarily divided age groups are represented in a series of histograms. The number of observed performances of each task, totaled through all of the age groups, are given on the left. The age groups (I–VI) and the average age of the workers in each group are listed at the bottom. The histograms are classified into three groups *A*, *B*, *C*, which are then identified as the temporal "castes." (From Wilson, 1976a.)

in this species, but the behavioral categories followed through time were too few to be certain.

The observed discretization in *Pheidole dentata* appears to represent an adaptation that increases spatial efficiency. It is obviously more efficient for a particular ant grooming a larva to regurgitate to it as well. Similarly, a worker standing "guard" at the nest entrance can be expected to be especially prone to excavate soil when the entrance is buried. The other juxtapositions in the *P. dentata* repertory make equal sense when the spatial arrangement of the colony as a whole is considered. The queen, eggs, first instar larvae (microlarvae), and pupae are typically clustered together and apart from the older larvae, although the positions are being constantly shifted, and pupae are often segregated for varying periods of time well away from other immature stages. Thus, the A-ensemble of workers can nicely care for all of these groups, moving from egg to pupa to queen with a minimum of travel. The mean free path of a patrolling worker is minimized by such versatility so that the least amounts of time and energy are consumed. It makes equal sense for A workers to assist the eclosion of adults from pupae, since pupae are already under their care.

It will be of interest to learn to what extent and by which patterns the temporal polyethism of other species of social insects has been discretized. We suspect that the recognizable age ensembles—that is, temporal castes—will prove to be related functionally to nest architecture. Species with complex nest structure, which provide a wider array of housekeeping tasks, and the opportunity for a more precise distribution of brood stages into chambers with differing microclimates, can be expected to have more temporal castes than those with relatively simple structures. For example, leafcutter ants of the genus *Atta*, which construct elaborate chambers for the gardening of symbiotic fungi, are expected to possess a correspondingly complex system of temporal polyethism. The patterns of temporal polyethism may be related to the dietary specialization of the species and the external environment of its nests. An ant species that forages widely outside the nest for small particles of food, while simultaneously defending its nest entrances from frequent attacks by predators, is likely to have an unusually discrete polyethic division at the end of worker life. The *C* age ensemble of *Pheidole dentata* represents just such a case.

To the extent that temporal castes are discretized, all of the castes of a species can be counted. *Pheidole dentata*, for example, has five adult castes: three temporal stages of the minor worker, a major worker (which has only one temporal stage), and the queen. The male does not occupy any known labor role. The larva might provide gluconeogenesis or some other metabolic service and hence constitute a sixth caste, but this possibility remains uninvestigated in *Pheidole*. There is every reason to suspect that the modest caste system of *P. dentata* is typical for the majority of ant species.

More generally, the total characterization of an insect society appears more feasible than it did only a few years ago. It is likely that the number of physical and temporal castes will not exceed 10 in ants and 20 in termites. The categories of behavior recorded in individual physical castes of ants have so far ranged between 20 and 42, with a broad overlap of categories among castes and a total species repertory probably not much exceeding 50. The number of categories of signals used in communication, mostly chemical, is likely to fall between 10 and 20 (Wilson, 1972, 1977). The sociograms based on such enumerations are far more feasible for insects than for primates and other higher vertebrates.

Since the work of Nolan (1924) and Rösch (1930) on honeybees, it has been repeatedly established that the temporal castes of social insects are to some extent flexible in behavior. To take one of the best-documented cases, when the number of wax-producing bees is reduced in a hive, some of the older workers reactivate their wax glands and recommence comb building. Similarly, if all younger bees are removed, including

149

those functioning as nurses, some of the older forager bees regenerate their hypopharyngeal glands and resume care of the larvae. This flexibility is nevertheless far from total. Older bees cannot regenerate their hypopharyngeal glands if pollen is lacking in their diet (Kratky, 1931). It has also been our experience that when older workers of *Pheidole dentata* are removed in large numbers from laboratory colonies, they are replaced as foragers only to a limited degree by the available force of younger workers, almost all of which remain in the nest with the brood and queen.

The existence of flexibility raises a semantic problem: why refer to age groups as castes if their labor roles can be altered according to the needs of the colony? The answer is that workers are not uniform in this regard. Probably each age group has a greater or lesser capability than others of changing in certain directions. In *Pheidole dentata*, for example, middle-aged, *B* workers are better able to shift to the roles of the oldest (*C*) workers than are the youngest (*A*) workers. Thus, a temporal caste is to be defined not just in terms of its labor profile within a normally constituted colony but also by its pattern of labor change when the age profile of the colony is altered.

The limited available evidence (see Wilson, 1971:136–196) shows that some species of social insects have a rigid temporal caste structure, others a far more flexible schedule of role change in the face of stress. The evolution of such patterns of response, and of their possible social and environmental correlates, remains wholly unstudied.

The greater flexibility of temporal castes might help to explain their near universal occurrence in social insects. They stand in contrast to physical subcastes of the worker caste, which are almost entirely limited to termites and ants and are few in number in any one species. Physical castes are much less easily shifted from one role to another not only in anatomy but in brain structure, sensory physiology, and innate behavior patterns (see, for example, Hecker, 1966; Goll, 1967; Wehner, 1969; Topoff et al., 1972).

150

4.6. THE NONEXISTENCE OF TEAMS

The relation of the members of an insect colony to one another can be characterized as one of impersonal intimacy. With the exception of the dominance orders of more primitively organized forms such as paper wasps and bumble bees, eusocial insects do not appear to recognize one another as individuals. Their classificatory ability is limited to the discrimination of nestmates from aliens, members of one caste as opposed to another, and the various growth stages among immature nestmates (Wilson, 1974b).

A consequence of this lower grade of discrimination is that members of colonies do not form cliques and teams. Groups assemble to capture prey, excavate soil, and other functions requiring mass action; and odor trails and other sophisticated techniques have evolved that permit the rapid recruitment of nestmates to the work sites. But the participants are entirely interchangeable. There is no evidence that they come and go in teams.

The lack of team organization is not necessarily the outcome of the limited brain power of social insects. It can be shown that at a very general level (see Figure 1.4 above) processes are less efficient when conducted by redundant teams than when conducted by redundant parts not organized into teams. This disparity can be overcome or reversed, as in fact it is in human beings, only if the degree of coordination among the members of the teams or between the teams is sufficiently great to compensate for the shortcomings inherent in the system redundancy.

4.7. WITHIN-CASTE ELITISM, SPECIALIZATION, AND IDIOSYNCRASY

Workers of some ant species, especially those belonging to the phylogenetically advanced subfamilies Dolichoderinae and Formicinae, vary greatly in their readiness to work. In the original study of this phenomenon, Chen (1937) found that "leader" workers of the carpenter ant *Camponotus japonicus* "subspecies *aterrimus*" begin to dig earth sooner when placed in earth-filled jars, move more earth per individual, and show less variation in effort than others. They furthermore have a stimulating effect on their more sluggish nestmates. Similar "elitism" has been observed in *Tapinoma erraticum* during brood transport (Meudec, 1973) and *Formica fusca, F. sanguinea* and *Camponotus sericeus* during adult transport (Möglich and Hölldobler, 1974, 1975). When Möglich and Hölldobler removed the small fraction of workers transporting their nestmates during a change in nest site, the emigration virtually ceased.

Without further studies across several behavioral categories, it is impossible to say whether the most active ants in one category are generally elites or mere specialists in the category under observation. A striking degree of specialization has been recorded in some formicine ants, especially in wood ants of the *Formica rufa* species complex. Horstmann (1973), for example, observed that foraging workers of *Formica polyctena* fall into one or the other of three kinds: arboreal foragers that collect honeydew primarily and search for prey secondarily, ground foragers that hunt prey almost exclusively, and collectors of nest materials. Individual ants remained in one category or another for periods of at least two weeks. Within-caste specialization may even be idiosyncratic in degree. Workers of *Lasius fuliginosus* patrol certain portions of the foraging ground over periods of weeks or longer, during which time they become familiar with specific portions of the terrain, the phenomenon of *Ortstreue* (Dobrzańska, 1966). Similar learning and particularization of behavior has been noted in the leafcutter ants of the genus *Atta*, harvester ants of the genus *Pogonomyrmex*, wood ants of the *Formica rufa* group, and honeybees (Jander, 1957; von Frisch, 1967; Rosengren, 1971; Lewis et al., 1974; Hölldobler, 1976; Herbers, 1977).

Not all apparent elitism can be explained away as specialization, idiosyncracy, and *Ortstreue*. In the detailed protocols published by Otto (1958) on *Formica polyctena*, it is apparent that a few workers were much more active than others in the pursuit of a multiplicity of tasks. Furthermore, some were more catholic in their choice of roles over most or all of the adult life span. Hence substantial variation can exist among individual colony members within the broad role sectors of particular age-size classes.

Very little is known concerning the basis of elitism but it appears that both innate and learned components can be important. Bernstein and Bernstein (1969) found that the ability to run a maze by *Formica rufa* workers is positively correlated with the size of the head, the diameters of the compound eyes and median ocellus, and the dimensions of the calyxes of the corpora pedunculata, the latter structures being the part of the brain most conclusively implicated in the control of complex behavior. Whether this variation is genetic or merely the outcome of random developmental variation is not known. The important point is that in either case it is fixed at the beginning of the adult instar. Changes induced by experience can also be major and long-lasting: when Jaisson (1975) prevented adult *F. polyctena* workers from contacting cocoons during the first fifteen days following eclosion, they proved incapable of tending cocoons in later life.

To summarize, it is clear that individual ant workers do not fit the popular image of invariant replicas that perform like parts in a machine. Some individual peculiarities occur that are based simply on the learning of specific portions of the environment, an obvious means of improving efficiency. But other, often striking deviations exist that are something of a mystery. They are either accidents of development and learning—the irreducible noise in the replication process—or else represent an adaptive form of diversification by which the colony divides labor still more finely beyond that attained through the physical and temporal castes. This distinction should become clearer and more significant as studies of

153

development are pursued and detailed properties of variation are incorporated into ergonomics models.

SUMMARY

Insect colonies perform twenty or more easily distinguishable behavioral acts in their daily repertories. Sets of acts connected by high transition probabilities can be conveniently defined as roles, even when on occasion the acts have entirely different functions. Sets of colony members that specialize on one or more roles constitute a caste (Figure 4.1). A task is used to denote a particular sequence of acts that accomplishes a specific function, such as nest repair or defense against an enemy. Individual tasks are usually either identical with individual roles or else consist of subsets of the acts that constitute individual roles.

With the exception of two cases of marginal significance, the castes of all species of social insects thus far studied are physiologically rather than genetically determined. Ants are limited in the number of castes generated by this means. The size-frequency curves of polymorphic species are mostly unimodal or bimodal, and only rarely trimodal, while their allometric growth curves (which determine the proportions of body parts) typically have a single slope and in only a small minority of species contain as many as three or four segments with differing slopes. All of these effects can be generated by a small set of very simple transformational rules that operate during larval growth and adult development within the pupa (Figures 4.6 through 4.8). A corollary result is that the evolution of caste systems in ants as a whole appears to consist of a relatively few steps (Figure 4.5). Physical castes in termites are based on two or three diverging pathways in hemimetabolous development, with each instar having the potential of serving as a distinct caste. In spite of their different physiological provenance, the caste systems are comparable in complexity and function to those of ants.

The proportions of castes are fine-tuned in each colony by the operation of as many as five caste-determining factors, among which nutrition and inhibitory pheromones are especially prominent. Some of these agents operate as negative-feedback controls, while others are token stimuli by which the physical environment is monitored. The particular sets adopted vary greatly among species. The overall pattern thus formed has yet to be explained in terms of adaptation to the environment.

Temporal castes, unlike physical castes, are almost universal in the social insects. In theory at least, temporal castes can be based on changes in responsiveness that are out of phase with reference to different tasks, creating a "continuous" system; or they can be based on sets of simultaneous changes, resulting in a "discretized" system (Figure 4.9). An example of a dis-cretized caste system is given in this chapter, and its apparent significance in colonial organization is explained.

Workers of insect colonies do not recognize each other as individuals, and hence the work forces are not organized into teams. Yet this can be shown to be a better strategy. In simpler processes, greater efficiency is obtained if all of the engaged workers act as interchangeable parts.

Division of labor has been elaborated in many species by elitism, defined as the existence of exceptionally active or entre-preneurial individuals within age-size cohorts. It has been further increased by other forms of idiosyncratic and specialized behavior, including *Ortstreue*, the fidelity of individuals to particular sectors of the foraging area. These modes of within-caste diversification vary greatly in kind and degree among species, a subject which will be taken up again in Chapter Seven.

CHAPTER FIVE

Caste Structure and Ergonomic Optimization

The next three chapters present a general conceptual framework around which caste studies can be planned. In this chapter we first introduce the definitions necessary to construct a quantitative theory, then provide a qualitative discussion of the major determinants of caste evolution. Chapter Six contains a number of mathematical models that address various aspects of caste determination. Chapter Seven forges the link between the demographic-level models and individual insect behavior.

Caste evolution has been guided by conflicting selective forces that promote ergonomic efficiency on the one hand and risk avoidance on the other. We can therefore usefully classify the life styles of eusocial species by the ways in which the compromise between return and risk is struck. For the most part attention will be focused on species that exhibit marked physical caste polymorphism. In order to start the investigation we pose the following question: given a fixed set of environmental challenges and opportunities, and a species equipped with an infinitely flexible genotype, what is the optimal compromise solution?

5.1. THE CASTE DISTRIBUTION FUNCTION

In Chapter Four it was argued that the fundamental property of a caste is its functional role in the colony's operation. Since this is a behavioral definition that is difficult to implement operationally, we introduced a number of surrogates which could serve as observable features of caste structure. They were:

156

(1) chronological age, a, which defined temporal caste cohorts; (2) a set of allometric size measurements, $\mathbf{s} = (s_1, s_2, \ldots, s_k)$, which identified physical castes by their allometric proportions; (3) activity level, x, which measures the "tempo" with which activities are performed. Other caste markers will surely be identified during future studies, and they can be added without great difficulty to the formalism.

If $(a, \mathbf{s}, x)$ is regarded as a complete specification of a caste, the caste structure of an entire colony is defined by the *caste distribution function* (CDF):

$$n(t, a, \mathbf{s}, x) \, da \, d\mathbf{s} \, dx = \text{number of individuals at time } t \text{ of age}$$
$$\text{class } [a, a + da], \text{ size class } [\mathbf{s}, \mathbf{s} + d\mathbf{s}],$$
$$\text{and activity class } [x, x + dx]. \qquad (5.1)$$

This description is intended to apply only to the worker castes. Reproductive castes are subject to an entirely different set of selective influences, some of which were discussed in Chapter Three. The total colony size is obtained by summing overall age, size, and activity classes:

$$\mathcal{N}(t) = \int_0^\infty \int_0^\infty \int_0^\infty n(t, a, \mathbf{s}, x) \, da \, d\mathbf{s} \, dx. \qquad (5.2)$$

Our working hypothesis will be that natural selection has shaped the CDF so that it is the optimum for the environment in which the breeding population of colonies lives. Thus we can expect a colony to be near the caste structure, n^*, that maximizes the number of daughter colonies in the next generation. Clearly n^* cannot represent an adaptive optimum in an absolute sense. Rather, it reflects the selection pressures of the local environment in which the species has existed in the recent evolutionary past. Furthermore, n^* is constrained by cost and risk considerations peculiar to the niche of the species.

In theory at least, the optimum caste distribution can be found by writing an equation of motion for the CDF and then maximizing the inclusive fitness of the colony with the CDF equation employed as a constraint. In practice, however, this is not possible, forcing reliance on simpler, approximate descriptions. In Appendix 5.1 the balance equation governing the CDF is derived. Many subsequent models will be special cases of this general equation. For the most part we shall deal with reduced descriptions of the CDF obtained by taking moments. For example:

$$n_a(t,a) = \text{age density} = \int\int n \, d\mathbf{s} \, dx$$

$$n_s(t,\mathbf{s}) = \text{size density} = \int\int n \, da \, dx$$

$$\bar{s} = \text{mean size} = \frac{1}{N} \int s n_s \, ds$$

$$\sigma_s^2 = \text{size variance} = \frac{1}{N} \int (s - \bar{s})^2 n_s \, ds$$

and so forth.

Since all of these quantities are changing in time, it is necessary to select a time scale of interest. Ultimately, our concern must be with changes in evolutionary time. However, the particular selective forces that shape caste evolution act on a much shorter time scale. The principal fluctuations during which selective forces can change are the following: (1) variations over geological time; (2) variations over the colony life cycle (which can be many years); (3) seasonal fluctuations; and (4) short-term changes concomitant with a sudden catastrophe such as a flood or predator attack. Thus, the time course of a caste variable might follow a complex pattern of the kind shown in Figure 5.1. Throughout our discussion the caste variables are assumed to be averaged over all time scales shorter than the one being employed. That is, $\bar{s}$ will denote $\frac{1}{T} \int_0^T s(t) \, dt$, where T is the time span of interest, either (1), (2), (3) or (4), above.

158

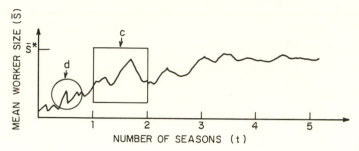

FIGURE 5.1. Variations in caste variables, such as mean worker size, $\bar{s}(t)$, must be considered with reference to widely disparate time scales.

If the CDF of a species is truly adaptive it will change as the individual colony ages, since the selective forces act differently on young as opposed to mature colonies. We must therefore include in the specification of the CDF the age of the colony from which it is drawn, that is, $n(t,a,\mathbf{s},x;A)$, where A is the colony age. In any community the population of colonies is not expected to be homogeneous in age. Instead, there will be a density distribution of colonies:

$C(t,A,\mathcal{N})$ = number of colonies at time t of age A and size $\mathcal{N}$.

At the level of colony selection, $C(t,A,\mathcal{N})$ is expected to be governed by a set of demographic equations of the same form as those governing the CDF for each colony. These are derived in Appendix 5.2. Throughout most of the current treatment we will neglect this complication and assume that the CDF is drawn from a population of colonies uniform in size and age.

5.2. ADAPTIVE DEMOGRAPHY

If the worker caste has indeed been shaped by colony-level selection, we can reasonably infer that the age and size-frequency distributions of workers are adaptive. This notion is best understood by an examination of age/size-frequency distributions of the kind displayed in Figure 5.2. The distributions

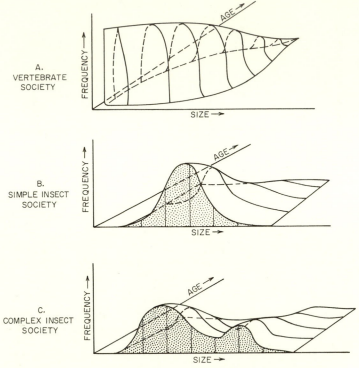

FIGURE 5.2. In social insects the externally defining qualities of caste are age and size rather than endogamously preserved genetic traits. Demography is therefore directly adaptive. This diagram illustrates the basic difference between the adaptive demographies of advanced social insect species and the nonadaptive demographies of vertebrate species. The examples depicted are based on the known general properties of real species but the details are imaginary. *A*: The distribution of the "vertebrate society" is nonadaptive at the group level and therefore essentially the same as that found in local populations of otherwise similar, nonsocial species. In this particular case the individuals are shown to be growing continuously throughout their lives, and mortality rates change only slightly with age. *B*: The "simple insect society," such as that of a primitively eusocial halictine bee, is subject to selection at the group level, but the age-size distribution of the worker caste does not yet show the effect and is therefore still close to the distribution of an otherwise similar, nonsocial population. The ages shown apply only to the imago, or adult instar, during which most or all of the labor is performed for the colony; and no further increase in size occurs with aging. *C*: The "complex insect society" has a strongly adaptive demography reflected in its complex age-size curve; in this particular case there are two distinct size classes, with the larger being longer lived. (From Wilson, 1975a.)

found in vertebrate societies are not basically different from those in populations of nonsocial vertebrates. They are epiphenomena that emerge from the schedules of individual mortality and natality, which themselves are shaped by natural selection at the level of the individual and its most immediate family. The array of age and size classes, in other words, is not "designed" to promote the welfare of the vertebrate society as a whole, or even that of one age-size group opposed to another. However, such design is manifest in advanced insect societies. Caste members are programmed to behave altruistically. They surrender most or all of their personal reproductive capacity in favor of the mother queen, undertake risky foraging trips, and sometimes literally throw their lives away in frenzied defense of the nest. What matters is not their personal survival and reproduction, but rather that of the queen.

Nonadaptive demography of the vertebrate type can be grasped largely from a study of the behavior and life cycles of individuals, but the adaptive demography of insect colonies must be analyzed holistically before the behavior and life cycles of the individual members take on meaning.

The existence of adaptive demographies has extensive consequences that deserve further empirical investigation. For example, one current explanation of senescence in organisms entails pleiotropy in the genes affecting the process of aging. Senescence is postulated to be under the control of genes that provide vigor or other favorable traits early in life but are responsible for physiological degeneration later in life (see Medawar, 1957; Williams, 1957). If the great majority of individuals die before age t due to natural, external causes such as predation, disease, and physical accidents, then senescence occurring from the age of t onward will have little influence on the contributions of individuals to the next generation. As a consequence, genes that provide advantages prior to age t but senescent degeneration from t onward will be favored by natural selection. In a free-living population one can expect to find many individuals at the peak of physiological vigor but few dying of old age. On the other hand, when a population of the same species is protected in a laboratory environment, death should occur primarily by senescence and at an age just beyond that attained by most individuals in the free-living state.

This theory applies with equal logic to adaptive demography and might be even more persuasively tested with examples from the social insects. When members of a particular caste are subject to a high accidental death rate, for example through predation during foraging trips, we should expect to encounter a relatively early senescence in members of the caste protected within a predator-free laboratory environment. In a word, altruism attracts senescence in the course of evolution. The survivorship curve of protected individuals should descend shortly after that of free-living individuals, and the principal cause of mortality will be senescent degeneration. (For the manifold characteristics of insect senescence, including especially brain degeneration, see Miquel, 1971.) Just such a relation appears to exist in honeybee workers, most of which are killed off during foraging by the age of 40 days and virtually all of which die of senescence by the age of 60 days (Rockstein, 1950; Sakagami and Fukuda, 1968).

Hence the theory predicts that castes facing the greatest risk of accidental mortality also have the earliest senescence. The survivorship curves obtained in the laboratory, now caused largely by senescence, will reflect the survivorship curves in the field. The latter are likely to be based mostly on accidental mortality and to descend at a somewhat earlier age. It is possible to be a little more specific. Minor workers should in general have the shorter survivorship schedules, since when older they forage almost daily, whereas major workers leave the nest less frequently and even when called out to defend the colony do not on most occasions suffer heavy mortality. Furthermore, in species where the major workers almost never go outside the nest, such as ants of the genera *Camponotus* (*Colobopsis*)

162

and *Zacryptocerus*, natural life spans should prove to be the longest of all. This hypothesis can be tested by seeing if there exists a correlation among species of ants between the percentage of time spent by major workers within the nest and the natural life span of members of this caste.

Still another prediction is that the very largest castes—the most massive major workers—will be the longest lived in both the free-living condition and the laboratory. This follows, first, from the fact that the colony has made a relatively large energetic investment in each individual, so that members of giant castes will be programmed to avoid accidental death, either through cowardice, or exceptional skill in dealing with predators, or both. Second, as a consequence, such individuals should be simultaneously programmed to delay physiological senescence.

However, even if it were true that the shape of the CDF has been molded by natural selection acting on such qualities as senescence, the degree of rigidity in the canalization of the caste proportions remains unknown. Consider the dimorphic ant species *Pheidole dentata*. Its major workers comprise from 3 to 52 percent of the adult population in wild colonies, with the proportion being between 3 and 12 percent in the great majority of cases.[1] Does this observed variation in caste ratio among colonies reflect mere statistical variation around a single genetically fixed ratio characteristic of the species? Or can each colony adjust its CDF so as to match the frequency and intensity of environmental opportunities and challenges? In other words, can the CDF be adjusted on a demographic time scale to the colony's advantage?

The following answer would be the one most consistent with current theory. The caste ratio should be fixed, in the event that either (1) the intensity of the selective forces is constant through time or (2) the time scale of its fluctuations is shorter than the average interval between the moment of caste determination and adult eclosion during individual development. If, on the

163

other hand, the selective forces vary substantially over a period greater than the worker generation time, natural selection should favor a flexible caste ratio.

This rule of caste-ratio flexibility should apply to a colony at or near mature size. A very small, rapidly growing colony faces a different investment prospectus. When small colonies encounter contingencies at frequent intervals (especially assault by enemies), it pays to produce majors at the beginning. But when such contingencies occur infrequently, it will probably pay to defer major production in order to yield the largest number of minors possible in the shortest time and thus to maintain the maximum possible colony growth through the early, exponential phase. In other words, small colonies of the second kind of species gamble that they will make it through the exponential growth phase without a catastrophe such as an invasion by a larger colony of ants. If the catastrophe occurs, a small group of soldiers could not prevent it—so why waste energy and time making the soldiers? It is true that many ant species, for example those in *Atta*, *Pheidole*, and *Camponotus*, do defer substantial soldier production until later stages of colony growth. On the other hand, this does not appear to be the case in termites (Haverty et al., 1977).

5.3. CASTES AND TASKS: ALLOMETRIC SPACE AND THE ERGONOMIC SCALE

A species of social insect adapted to a particular ecological community specializes on a *microhabitat*. For example, it might (1) restrict its activities to large logs in intermediate stages of decomposition; or (2) excavate nests in the soil and forage through the surrounding leaf litter; or (3) nest in hollow twigs in trees and forage through the canopy. The chosen microhabitat presents the species with an array of *tasks* it must perform which are peculiar to its life style: it has access to a particular ensemble of prey and other food items, it confronts certain predators and competing species, and it has peculiar requirements for nest construction and brood care.

If the CDF is to be analyzed as the supposed adaptive reflection of the species' life style, then a way must be found to relate the function to the bewildering array of contingencies faced by colonies on a day-to-day basis. That is, the notion of task must be quantified just as caste was quantified earlier. One method is to employ the notion of an "ideal allometric design" suited for each task.

Recall that a caste is defined as a particular set of observable characteristics: (a) a "dynamic" specification of age, a, and activity, x; and (b) a "static" specification $\mathbf{s} = (s_1, s_2, \ldots s_k)$ consisting of a number of allometric measurements that provide the size and proportions of the individual workers. The points $\mathbf{s}$ in this k-dimensional *allometric space* represent what we mean by the physical castes. Now consider a particular task, such as foraging or carrying microlarvae. It is possible to imagine a worker whose allometric measurements, represented by a point $\mathbf{s}_0$ in allometric space, are "perfectly" matched to this task. Furthermore, the task itself can be specified by the point s_0, even in the absence of such a "perfectly designed" caste. In other words, a *physical task* can be represented as a point in the allometric space corresponding to the set of allometric measurements of the worker ideally suited for the task.

There is obviously more to a task than the physical properties required to perform it, and we shall expand our definition later to incorporate the notion of a time and energy budget to characterize a task as a sequence of behavioral acts. The allometric characterization of a task is meant only to specify the *physical* adaptations suited for its performance. Figure 5.3 is a schematic representation of one conceivable pair of caste and task distributions in two-dimensional allometric space.

In general, the species cannot accommodate the task array with uniform efficiency. The members of a colony are limited to a few growth stages and a relatively narrow range of allometric

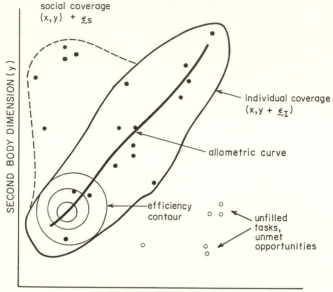

FIRST BODY DIMENSION (x)

FIGURE 5.3. The concept of the allometric space. Each microhabitat offers social insects a potential task array. Each task can be most efficiently dealt with by individual colony members of a given body form x, y. A species adapts by evolving a system of physical castes that can be specified with a tightly correlated allometric array, shown here as a heavy curve. Efficiency contours have been drawn around one point on the allometry curve, which represents a single caste; each caste performs tasks with declining efficiency the farther they are from it on the allometric space. Some task points are close enough to the allometric curve to be dealt with by individual colony members (individual coverage); others can be accommodated only through cooperative efforts (social coverage). Still others cannot be met at all; by definition these lie outside the niche of the species.

variation within each growth stage. Moreover, in the social Hymenoptera virtually all castes are drawn from the single adult instar. In the course of its evolution, the social species evolves one or more allometric arrays that wind through the allometric space. Depending on this deployment, its colonies are able to meet certain contingencies with great efficiency, others poorly, and some not at all. Each allometric variant (x, y)

in the colony has an individual coverage of tasks; it can perform not only the task closest to it in allometric space but perhaps others in the near vicinity. Hence its *individual coverage* can be defined on the space as $(x,y) + \varepsilon_I$, where ε_I is the individual caste's radius of performance. Each variant also has the potential of expanding its competence through communication and cooperation with its nestmates. This additional coverage is the *social coverage*, defined as $(x,y) + \varepsilon_s$. The farther the task from the position (x,y) of the variant, the less is the variant's efficiency likely to be. This particular decline in efficiency can be characterized as a system of contours. Individual coverage depends on behavioral flexibility, while social coverage depends in addition on the mechanisms of communication and cooperation. Thus, the task performance radii are not yet precisely defined quantities. However, as will be demonstrated in Chapter Six, the radii can be related to the ergonomic efficiency of each caste. Deferring for the moment the question of how the performance radii are determined, the schema of Figure 5.3 suggests how we can form a preliminary picture of the physical niche occupied by each species, based on its CDF. This perception will be expanded in Chapter Six in order to obtain an upper bound on the number of distinct castes required for a given task array.

5.4. THE RESOURCE DISTRIBUTION FUNCTION AND FORAGING EFFICIENCY

By employing the concept of allometric space it is possible to quantify the notion of physical castes and tasks. One of the principal task ensembles that has shaped the CDF is foraging. We can idealize the situation of a colony with respect to resource abundance as follows. A colony is furnished with a CDF, $n(t,a,\mathbf{s},x)$. The environment provides a stream of food items in the vicinity of the colony, comprised of such categories as arthropod prey, seeds, and droplets of aphid honeydew. This resource stream possesses a density function analogous to the

CDF. A food item can be characterized by its size dimensions, $\hat{\mathbf{s}}$, its age, $\hat{a}$, and any other relevant characteristics, $\hat{\mathbf{y}}$, such as its chemical composition and caloric content. Thus, we can define a *resource distribution function* (RDF), $R(t,\hat{a},\hat{\mathbf{s}},\hat{\mathbf{y}})$, where

$$R(t,\hat{a},\hat{\mathbf{s}},\hat{\mathbf{y}}) = \text{number of food items per unit}$$
$$\text{area at time } t \text{ of age } \hat{a}, \text{ size } \hat{\mathbf{s}}$$
$$\text{and chemical value, } \hat{\mathbf{y}}.$$

(The circumflex $\hat{}$ is used to distinguish quantities in the RDF as opposed to those in the CDF.) The age $\hat{a}$ of a food particle—that is, the time since it became available in the colony's neighborhood—is important when food items have a limited duration due to decomposition or to the harvesting activities of competitors.

This additional definition allows a more precise reformulation of the basic question underlying the ergonomic optimization theory of caste evolution. We ask, *what is the optimal CDF for a given RDF?*

In order to proceed, however, the connection between the two distributions must be defined. This connection is provided by the *foraging efficiency function* (FEF), η, which measures the per-capita resource utilization per unit time. The FEF depends on all of the CDF and RDF variables, i.e., $\eta = \eta(t,a,\mathbf{s},x;\hat{a},\hat{\mathbf{s}},\hat{\mathbf{y}};n,R)$, but in practice we will be forced to restrict our attention to special cases. For example, $\eta(s,\hat{s})$ is defined as $\eta(\mathbf{s},\hat{s}) =$ rate at which a forager of size s harvests food particles of size $\hat{s}$ (measured, say, in calories/unit time). This rate might well depend on the number of other foragers that are in the vicinity and available for a cooperative effort. Until specific models for η are formulated (see Chapter Seven) we shall continue to treat η phenomenologically. The total foraging return rate for the colony can be written

$$R(t) = \iint\limits_{0}^{\infty} \eta(s,\hat{s})n(t,s)R(t,\hat{s}) \; ds \; d\hat{s}.$$

This is essentially the "return function" employed in the reproductive strategy model of Chapter Two.

We hypothesize that the FEF has been optimized in evolutionary time by selective forces acting on behavioral and physical caste properties. Before addressing this process, however, it is necessary to face the issue of whether ergonomic efficiency is indeed a good surrogate for inclusive fitness with reference to caste evolution.

5.5. ENERGY AS A MEASURE OF UTILITY

The ultimate currency in colony-level natural selection is the summed production of reproductives over a colony generation, consistent with an optimal sex ratio which depends on the sex ratio in the population of colonies as a whole (see Chapter Three). In general, colonies appear to build as large a population of workers as possible during the ergonomic phase, then "cash in" with a partial or full conversion to reproductives during the breeding season. Consequently, a reasonable measure of fitness for a colony in the ergonomic phase is energy, with both the growth rate of the adult biomass converted to calories/unit-of-time and the full colony biomass achieved at the end of the ergonomic phase converted to calories.

The advantage of employing energy as the basic currency is that all colony processes can be converted into it, as follows:

(1) The biomass and biomass growth of the colony or any portion of it can be translated to energy units.

(2) Foraging success, brood care, and other nurturent activities can, in theory at least, be translated into energy gained.

(3) Protective activities, including nest defense and construction, can be converted into calorie equivalents saved because of the reduction in mortality and impaired colony functions that such activity avoids.

(4) Metabolism and mortality can be converted directly into caloric loss.

169

It is admittedly an oversimplification to measure ergonomic success entirely by the energy equivalents of biomass. Reproductive success depends not just on the colony size at the end of the ergonomic phase but also on the proportion of immature and mature forms as measured by the caste distribution function at this critical time. It also depends on the forms and kinds of "capital" at the disposal of the colony, particularly the amount of stored resources and the structure and location of the nest. Nevertheless, we will frequently regard energy equivalence as the probable overriding determinant of reproductive success. The optimal caste distribution function and nest structure at the onset of the reproductive phase are unlikely to differ from those built up during the ergonomic phase, while few species store large amounts of food materials near the end of the ergonomic phase.

However, it is important to bear in mind that maximizing ergonomic efficiency is generally not the same as maximizing inclusive fitness. The reason is that the conversion from energy to genes (i.e., to reproductive alates) is not linear. Indeed, during the ergonomic stage of colony growth the efficiency with which energetic reserves can be converted into live biomass might exhibit variously increasing, constant or decreasing returns to scale. The importance of this nonlinearity can be demonstrated by reexamining the reproductive strategy model employed in Chapter Two. There the role of ergonomic constraints was implicit in the "return function." If the model is expanded to include the colony's energetic reserves as an explicit state variable, the constraint equations can be written:

$$(1) \quad \frac{dW(t)}{dt} = M_W[uE, W] - \mu W \qquad W(0) = W_0$$

$$(2) \quad \frac{dQ(t)}{dt} = M_Q[vE, W] - vQ \qquad Q(0) = Q_0$$

(3) $\quad \dfrac{dE(t)}{dt} = \eta(W)W - m_W W - m_Q Q - M_W[\,\cdot\,] - M_Q[\,\cdot\,]$

$$E(0) = E_0$$

(4) $\qquad 0 \leqslant u + v \leqslant 1, \qquad W, Q, E \geqslant 0$

The structure of the model is shown in Figure 5.4. The term $\eta(W)$ is the per capita foraging efficiency, which will be developed more fully in Chapter Six; for the present η can be regarded as a random variable that models the effects of fluctuations in the foraging returns to the colony. The terms m_W and m_Q are the average per capita metabolic costs of the respective worker and queen castes, while μ and v are their mortality rates as before. The functions $M_W[\,\cdot\,]$ and $M_Q[\,\cdot\,]$ are the "manufacturing" rates for workers and queens, respectively.

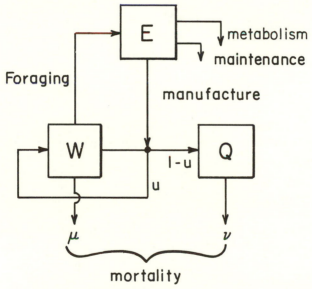

FIGURE 5.4. The energy reserves E of the colony are incorporated into the model of resource allocation.

In general, these quantities depend on the size of the worker force available for brood rearing and other duties, as well as on the fraction of resources allocated to producing workers (uE) and queens (vE). In Chapter Two we made the simplest assumption that $M_W[\cdot]$ and $M_Q[\cdot]$ were *linear* in uE and W: $M_W = (\text{const.}) \times uEW$, $M_Q = \text{const.} \times (1 - u)EW$, where the return function, $R(t)$, incorporates the time dependence of $E(t)$. This is not likely to be true as a rule; the effects of scale, diminishing returns, and so forth (cf., Chapter Two) will admit linearity only over a small operating range. Indeed, in Chapter Two we pointed out that nonlinearities in the energy utilization, uE, could result in "graded" reproductive strategies, with workers and alates produced simultaneously. Here we shall focus our attention on the consequences of nonlinearity coupled with ever present random disturbances.

Since η, M_1 and M_2 incorporate random fluctuations, Q, W and E are themselves random variables. Thus, the fitness criterion for the colony is

(5)
$$\underset{0 \leqslant u \leqslant 1}{\text{Max}} \; \langle Q(t) \rangle$$

where $\langle \cdot \rangle$ indicates the expected value, taken with respect to the probability distribution underlying $Q(t)$.[2] From the problem formulation it is clear that the fitness criterion (5) will yield a different strategy from maximizing either $\langle E(t) \rangle$ or the foraging efficiency over the season,

$$\left\langle \int_0^t \eta(w) W \, dt \right\rangle$$

For example, consider the static (single period) optimization of queen production, where $W = \text{constant}$ and $v = 0$. Then the relation between Q and E might have the form shown in Figure 5.5.[3] Now E is a random variable with mean $\langle E \rangle$ and variance σ_E^2. Therefore, expanding $Q(E)$ about $\langle E \rangle$ in a Taylor series.

172

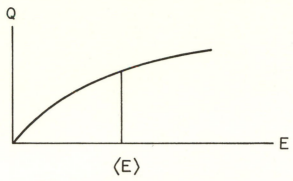

FIGURE 5.5. The relation between the colony fitness, measured by queen production Q, as a function of the energy resources E of the colony. In this imaginary example, $\langle E \rangle$ denotes the average energy reserves accumulated.

$$Q(E) = Q[\langle E \rangle] + (E - \langle E \rangle)Q'(\langle E \rangle)$$
$$+ \tfrac{1}{2}(E - \langle E \rangle)Q''(\langle E \rangle) + \cdots$$

Taking the expected value of both sides, we obtain

$$\langle Q(E) \rangle = Q(\langle E \rangle) + \tfrac{1}{2}\sigma_E^2 Q''(\langle E \rangle) + \cdots$$

Thus, maximizing the expected number of alates is seen to be equivalent to maximizing the expected energy only when $Q(E)$ is linear. Later, we shall use the variance, σ_E^2, as a measure of "risk" or environmental "uncertainty," and we shall see that uncertainty can profoundly alter caste ratios. In the dynamic model the variations in reproductive strategies engendered by maximizing inclusive fitness surrogates such as ergonomic efficiency can be quite large depending on the size of stochastic effects and the degree of nonlinearity in the manufacturing and foraging functions.

5.6. DEFENSIVE CONTINGENCIES: RISK VERSUS RETURN

The other major contingency that shapes the evolution of caste systems is the necessity for defending the colony from

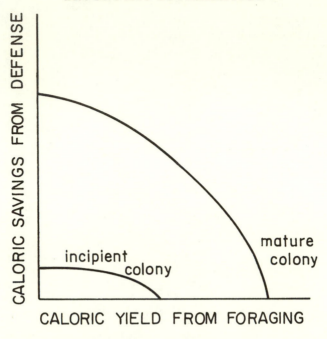

FIGURE 5.6. The tradeoffs in performance of two tasks by the colony as a whole. The curves of young colonies are expected to differ from those of mature colonies.

predation. In some cases this has resulted in such bizarre adaptations as the "door-head" major workers in the ant genus *Zacryptocerus* and termite genus *Cryptotermes*, as well as the soldiers of ant and termite species (Wilson, 1971; Maschwitz and Maschwitz, 1974) specialized to explode and spray sticky secretions on invaders. The particular form of each defensive adaptation may prove to be an evolutionary imponderable (see Chapter Eight). Nevertheless, from the ergonomic viewpoint one can measure defensive costs in calories by the concept of "opportunity loss." That is, investment in a soldier caste has two ergonomic consequences: (1) it decreases the size of the foraging work force, since soldiers seldom perform double duty

174

as foragers; and (2) it increases the colony reliability, that is, the probability of the colony surviving to the reproductive phase. While the former effect decreases the foraging return rate the latter increases the *expected* reproductive output. In addition to protecting the colony, soldiers often guard foraging columns and larger food items discovered by workers, thus increasing the colony's foraging effectiveness. An optimum tradeoff can therefore be expected between investment in the soldier and worker castes. In Figure 5.6 we have suggested how the form of this tradeoff might look. Note that the tradeoff curves of small or incipient colonies differ from those of large, mature colonies. In order to obtain the optimal worker-soldier caste ratio one would have to superimpose a fitness function onto the graphical analysis. In the next chapter we shall discuss how the situation can be modeled quantitatively.

SUMMARY

In this chapter an accounting system of castes and division of labor is devised that can be translated into relative genetic fitness. The principal measure is the *caste distribution function* (CDF):

$$n(t,a,\mathbf{s},x)$$

where in a colony at time t, n is the number of colony members in age group (temporal caste) a, size (physical caste) $\mathbf{s}$, and activity level x. The "goal" of the evolving species is postulated to be the CDF that attains a local maximum of colony genetic fitness. This local maximum is set in turn by the special conditions of the environment in which the colony lives.

In order to specify the nature of the microhabitat in which the colony lives, we introduce the concept of the *allometric space*: the set of physical measurements that is best suited for each of

175

the tasks presented to the colony by the microhabitat. We hypothesize that by means of natural selection the species evolves an allometric growth curve and modifies the flexibility of individual behavior in such a way as to provide an adequate coverage of the task points in the allometric space (Figure 5.3).

The structure of the allometric space allows us to define the *resource distribution function* (RDF), which is the relative frequency of different classes of available food items, classified according to the amount of time they have been available, their size, and their chemical properties. Also required is a *foraging efficiency function* (FEF), which specifies the rate and efficiency at which individual castes utilize particular resources.

Ultimately, therefore, the relation between the CDF and colony fitness (measured by the production of new reproductive individuals) is determined by the natality and mortality schedules of the physical castes, their efficiency, and the life span of the mother queen. In order to translate this relation into a form usable in quantitative analysis, a single utility measure is suggested: colony growth, efficiency, and mortality are all expressed in energy units. With a single utility measure it is possible to describe the optimization process as a search for the CDF that produces the maximum rate of colony growth under natural environmental conditions. However, only in rare cases will it be possible to treat the fitness of a colony as a linear function of its energy reserves.

APPENDIX 5.1. THE CDF EQUATION

The equation of motion governing the caste distribution function is derived by writing a balance law for the density of points $n(t,a,\mathbf{s},x)$ in the caste phase space. Consider the volume element $dV = da\,d\mathbf{s}\,dx$ located at point $(a,\mathbf{s},x)$ in caste space (see Figure 5.7). A conservation law for the phase points has the

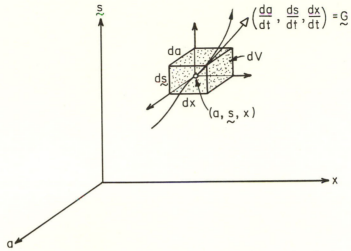

FIGURE 5.7. The balance equation for the caste distribution function is displayed here for the volume element dV in the caste phase space.

form:

$$\frac{dn}{dt} = -[\text{net efflux from } dV] \pm [\text{sources/sinks}]$$

$$= -V \cdot (Gn) \pm S,$$

where $\mathbf{G} = \left(\dfrac{da}{dt}, \dfrac{d\mathbf{s}}{dt}, \dfrac{dx}{dt}\right) = (1, \mathbf{g}_s, \mathbf{g}_x)$ is the growth velocity in age, size and activity. $\left(\dfrac{da}{dt} = 1\right.$ if age is measured chronologically.)

Phase points are lost via mortality and by transitions from activity x in dV to points x' outside of dV. Phase points are gained by birth of new individuals and by transitions into dV from points outside. (The term $g_x = \dfrac{dx}{dt}$ allows for continuous changes in tempo with age or size.) Thus, we can write the

177

balance equation as:

$$\frac{\partial n}{\partial t} = \underbrace{- \frac{\partial n}{\partial a}}_{\text{aging}} \underbrace{- \frac{\partial}{\partial \mathbf{s}}(\mathbf{g}_s n)}_{\text{size growth}} \underbrace{- \frac{\partial}{\partial x}(g_x n)}_{\text{activity growth}} \underbrace{- \mu n}_{\text{mortality}}$$

$$+ \underbrace{\int_{x\varepsilon dv} T(x,x')n(x')\,dx'}_{\substack{\text{sudden transitions} \\ \text{into } dV}} - \underbrace{\left[\int_{x\varepsilon dv} T(x',x)\,dx\right]n(x)}_{\substack{\text{sudden transitions} \\ \text{out of } dV}}$$

$$+ \underbrace{B(t,a,\mathbf{s},x)\delta(a)\delta(\mathbf{s} - \mathbf{s}_0)\delta(x)}_{\text{birthrate}}$$

APPENDIX 5.2. THE DEMOGRAPHIC EQUATIONS FOR COLONY-LEVEL SELECTION

The conservation equation governing the colony distribution function $C(t,A,\mathcal{N})$ giving the number of colonies in a community of age A and size $\mathcal{N}$ at time t is:

$$\frac{\partial C}{\partial t} + \frac{\partial C}{\partial A} + \frac{\partial}{\partial \mathcal{N}}[GC] = -\hat{\mu}C,$$

where

$$G(\cdot) = \frac{d\mathcal{N}}{dt} = \text{colony growth rate}$$

and

$$\hat{\mu}(\cdot) = \text{colony mortality rate.}$$

The birthrate of new colonies is given by

$$C(t,0,1) = \int\!\!\int_{0}^{\infty} C(t,A',\mathcal{N}')B(t,A',\mathcal{N}')\,dA'\,d\mathcal{N}',$$

where

$B(t,A'\mathcal{N}')$ is the per colony rate at which colonies of age A' and size $\mathcal{N}'$ give rise to new colonies at $A = 0$ and $\mathcal{N} = 1$ (i.e., the founding queen).

178

The form of these equations is nearly identical to the ordinary demographic equations. Therefore, at first glance it may appear that one could proceed to treat colonies as one treats individuals in a population and go on to develop competition theory analogous to the classical treatment. However, the situation is more complex in such colonial systems because the colony birth, growth, and mortality functions depend on the caste distributions within each colony.

NOTES TO CHAPTER 5

1. These estimates are based on censuses of free-living colonies made by E. O. Wilson and R. A. Metcalf in the vicinity of Tallahassee, Florida, during April 1975.

2. This fitness criterion is consistent with the inclusive fitness formula for colony-level selection already developed (Chapter Three) if the population size, colony size, and sex ratios are assumed to be stable:

$$V_c = \sum_1 r_{cj} S_j \langle N_j \rangle = \text{const.} \times \langle Q \rangle$$

3. An example of labor saturation has been experimentally documented by Brian (1953). He measured the number of surviving larvae produced from an egg brood as a function of the number of nurse workers attending them. His data fit a saturating curve, which indicates that the workers' efforts are nearly additive at low numbers but become almost inversely related to density at higher levels.

CHAPTER SIX

Optimum Caste Ratios

A complete theory of caste ergonomics will unite the entire caste and resource distribution functions by means of a comprehensive model for the foraging efficiency function. Because so much of the effort must be based on painstaking empirical studies, this goal is not yet within our grasp. What we can accomplish is the construction of reduced models that identify the main features of caste allocation.

In the present chapter we will attempt the crucial first step of explaining why the number of allometrically distinct castes is always considerably less than the number of castes. By this means we will justify limiting the initial models to a small number of allometric types. The smallness of caste numbers also becomes a focus of research in its own right: we are able to classify the various forces promoting and inhibiting caste proliferation and to devise a method for computing the optimal caste size distribution that corresponds to a given resource distribution. The models can then be made more realistic by adding the roles of environmental uncertainty and predation pressures that comprise the difficult-to-measure "risk" factors of colony ergonomics.

6.1. HOW MANY CASTES?

In principle, the number of distinct allometric castes is bounded by the number of distinct tasks that must be performed by the colony. There is no *a priori* reason to expect caste redundancy, in which two or more different allometric castes are specialized on a single task. Indeed, we shall demonstrate presently that if no constraints are imposed on caste evolution, the optimal caste number is just equal to the number of tasks.

The precise number of tasks occupying an insect colony is difficult to determine, but lower limits can easily be set. In a series of studies on ecologically disparate species in the ant genera *Leptothorax, Pheidole, Solenopsis, Cephalotes,* and *Zacryptocerus*, the total behavioral repertory of the worker caste has been estimated to lie between 20 and 45, with each species determined to a particular number within a 95 percent confidence interval of ± 3 or less (see Table 4.1 and Wilson, 1976a,b; Corn, 1976). Of these behavioral acts, only four can be classified as "personal" behavior with no apparent social function: feeding, self-grooming, excreting, and a peculiar activity called "antennal tipping." Thus we can reckon conservatively of the existence of at least 20 distinct tasks, and probably 35 or more tasks in some species, to which castes might be specialized.

In nature, we find that the number of castes in individual species of social insects lies far below the number of tasks. Of the 263 living ant genera listed by Brown (1973), no less than 209, or 80 percent, consist entirely of species that are monomorphic: their worker caste is not divided into anatomically distinct subcastes. Of the other genera, 40 contain at least some species polymorphic to the extent of having two recognizable castes; this category includes cases in which the two forms are disjunct and others in which they are connected by intermediates. We know of only 3 genera, *Atta, Daceton,* and *Eciton,* with three easily recognizable worker castes; to this group probably also belongs *Pheidologeton*. Another 3 ant genera consist entirely of workerless parasites, while 7 cannot be characterized at this time because of lack of adequate information (see Table 1.1).

Temporal castes are not likely to be much more numerous. As documented in Chapter Four, the minor workers of *Pheidole*

181

dentata go through three discrete age periods and the major workers only one. The total number of physical and temporal castes in this species is not likely to exceed six—less than one-fourth the number of tasks performed by the colony as a whole. The species of bees and wasps are virtually all monomorphic, with a temporal division roughly comparable in complexity to that of ants. Some termite species, particularly members of the phylogenetically advanced members of the family Termitidae, appear to have more complex caste systems. In these insects temporal stages and physical castes are nearly the same thing: beyond the initial larval stage, which is a period of helpless dependence on older nestmates, each individual passes through a series of instars during which it changes in anatomy and behavior. Furthermore, both females and males become workers, with sexual differences adding to the number of castes (Noirot, 1969). The total number of active instars, both male and female, is eight in *Amitermes* and *Acanthotermes* and twelve in *Trinervitermes*. To these stages might be added temporal division of labor *within* instars, a refinement described in *Nasutitermes* and *Zootermopsis* (Pasteels, 1965; Howse, 1968; Renoux, 1976). Hence it is conceivable that in extreme cases such as *Trinervitermes* the number of termite castes could indeed approach the number of tasks. This possible distinction between the social Hymenoptera and the higher termites deserves closer investigation.

The disparity between the number of tasks and the number of castes in social Hymenoptera and at least some species of termites reveals something important about the division of labor. Ants and termites have been present since mid-Cretaceous times, for about 100 million years, and modern polymorphic genera such as *Camponotus* and *Oecophylla* for at least 35 million years. It seems reasonable to conclude that existing caste systems represent equilibria, and to reject the hypothesis that castes are slowly multiplying toward a higher, more nearly optimum number close to or identical with the number of tasks.

182

We can use the allometric space concept to place an upper boundary on the caste number. Consider the simplest case in which $\mathcal{N}_T$ task points are distributed at random over the species' physical niche, which occupies an area, A, in allometric space. (The resulting Poisson distribution can be replaced by any clumped distribution, such as the one chosen for Figure 5.3, without affecting the qualitative outcome.) The probability that the number of tasks in a given local region of area a is equal to k is given by

$$\Pr\{\text{no. of tasks} = k\} = e^{-a\rho_T}\left[\frac{(a\rho_T)^k}{k!}\right] \qquad (6.1)$$

where ρ_T = average density of tasks = $\mathcal{N}_T/A$. Alternatively, we could consider a clumped distribution and redistribute the points into an equivalent random distribution. The result will be an overestimate of the caste density, ρ_C, required to handle the given task array ρ_T.

Over each task point in the allometric space sits an ergonomic efficiency function η, depicted in Figure 6.1. For simplicity we assumed that η depends only on the distance between the caste and the tasks and that there is a minimum performance for

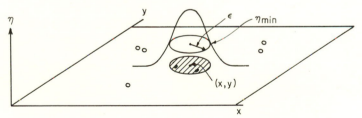

FIGURE 6.1. The efficiency function η of a single caste x,y is projected over the allometric space. The effectiveness of an average individual of type x,y in the performance of a given task falls off with the distance separating it from the caste best suited to perform the task. Beyond a certain distance ε, individuals cannot perform the task with the minimum level of efficiency required to make the effort energetically profitable.

183

each task below which the attempt is energetically unprofitable. This specification in turn establishes an effective minimum performance radius, ε, around each caste. Finally, it is assumed that the caste points (x, y) are distributed at random over the allometric niche.

Now the problem is reduced to the following question: what is the minimum caste number at which the essential tasks are covered by the discs of radius ε centered on each caste? To be somewhat more precise, if both castes and tasks are distributed randomly over the space A in allometric coordinates, with densities ρ_C and ρ_T respectively, and if the caste points are assigned an effective performance radius ε, then the quantity desired is the caste density required so that a randomly chosen task is "covered" by a caste with a probability of, say, α.

If both tasks and castes are Poisson-distributed over the allometric space, the probability that the distance from a task to the nearest caste is less than ε is

$$\Pr\{\text{task-to-caste distance} < \varepsilon | \rho_C\} = 1 - e^{-\pi \varepsilon^2 \rho_C}. \quad (6.2)$$

If N_T is the total number of tasks, of which we allow only n_0 to go unperformed by any colony member (i.e., $\alpha = 1 - n_0/N_T$), then

$$\{1 - [1 - e^{-\pi \varepsilon^2 \rho_C}]\} < \frac{n_0}{N_T} \quad (6.3)$$

or

$$N_T e^{-(\pi \varepsilon^2/A)N} \leqslant n_0$$

and

$$N_C \leqslant \frac{A}{\pi \varepsilon^2}[ln N_T - ln\, n_0]. \quad (6.4)$$

Thus, the number of castes, N_C, is bounded roughly by

$$N_C \gtrsim \frac{1}{\varepsilon^2} ln N_T. \quad (6.5)$$

The number of castes required for a prescribed performance level is therefore much smaller than the number of tasks. This reduction in caste number is likely to be somewhat weaker than logarithmic since neither ρ_C nor ρ_T are uniformly and randomly distributed over allometric space. In fact, a species can alter its allometric curve so as to approach higher concentrations of task points than would be feasible in a blind, unselected random array of caste points.

Another striking correlation emerges from the relation $\mathcal{N}_C \sim 1/\varepsilon^2 \, ln\mathcal{N}_T$, when the meaning of the task performance radius ε is examined more closely. The caste efficiency function η, on which ε depends, clearly must involve more than just the allometric specifications of the caste, just as manual dexterity in human beings is based on more than the anatomical dimensions of the hand. Behavioral flexibility and the capacity to cooperate can enormously increase the range of tasks to which an individual can contribute. Thus, increasing ε corresponds to increasing the behavioral flexibility of a caste. From the caste/task formula it is seen that the number of required castes decreases as $1/\varepsilon^2$. If we characterize each caste by $k > 2$ measurements on the allometric space, rather than just 2, $\mathcal{N}_C$ will vary approximately as $1/\varepsilon^k$. Thus, the need for distinct physical castes decreases strongly as the behavioral flexibility of each caste increases.

Another way of viewing this dependence is to plot as a function of ε the probability that the average separation between the tasks and castes is less than ε:

$$\Pr(d_{TC} < \varepsilon) = 1 - e^{-\pi\varepsilon^k \rho_c} \tag{6.6}$$

where k is the number of allometric measures required to specify a physical caste. Figure 6.2 presents the general curve for the case of $k > 2$. In general, the more specialized the physical castes, that is the higher the value of k, the more abruptly does the need for such physical castes disappear as behavioral flexibility (ε) increases. This creates an apparent paradox. How

185

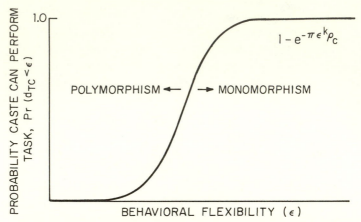

FIGURE 6.2. The probability that a caste can perform a task with adequate competence increases abruptly when the behavioral flexibility of the caste reaches a certain level. The shift from polymorphism to monomorphism can occur with a relatively small increment in either individual behavioral flexibility or the capacity to work cooperatively.

can physical caste specialization by itself lead to a more efficient coverage of the essential tasks, given a constant number of castes? The empirical answer seems to be that k and ε are inversely correlated: the more physically specialized an individual colony member, the less flexible is its behavior.

The cardinal inference is that for a given degree of physical specialization (k) a small change in behavioral flexibility can result in the shift in the colony optimum from monomorphism to polymorphism or back again. The phenomenon is reminiscent of "phase transition" curves in physics, which characterize sudden condensations, shifts from order to disorder, and other abrupt transitions. As suggested in Figure 6.3, the tradeoff between physical and behavioral castes is not proportional: the strong selective pressures for morphological differentiation should disappear rather quickly as behavioral complexity increases. The same is true as the ability of members of the same caste to cooperate increases, enlarging the social coverage of the allometric space.

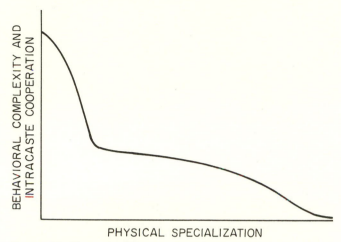

FIGURE 6.3. The tradeoff between physical castes and behavioral flexibility is unlikely to be proportional. As suggested in this curve, which is selected from many possible of the same general form, a phase transition exists in which the optimum strategy of a species can shift abruptly from multiple specialized castes to a single flexibile worker caste.

If an evolutionary "phase transition" exists, we should expect to find genera or other sets of phylogenetically closely related species that display striking variation in the degree of worker polymorphism. This is notably the case in ants. Of the 44 genera with polymorphic species, at least 11 also contain some monomorphic species (*Myrmecia, Tetraponera, Neivamyrmex, Monomorium, Solenopsis, Pogonomyrmex, Crematogaster, Strumigenys, Iridomyrmex, Azteca,* and *Formica*). Some other genera that are fully polymorphic nevertheless display strong interspecific variation in the degree of polymorphism. The species of *Camponotus*, for example, range over the entire gamut: from feeble monophasic allometry to complete dimorphism.

We envisage three means by which the transition occurs in the direction of monomorphism. First, the behavior of workers in particular age-size cohorts can simply become more complex and flexible. The second means is to add temporal castes: the members of a given size cohort specialize according to age. This trend, which is followed by virtually all of the higher social insects, has the advantage of allowing reversible specialization. Individuals are not frozen within any particular anatomy or brain circuitry. When the needs of the colony demand it, workers can switch to a new behavioral regime and even, given a few days time, reactivate dormant exocrine glands. Finally, members of the same age-size cohort can add to their repertory by cooperative action—working together to move large prey items, defending the nest against enemies too formidable for a single member, and similar actions. This increment is the "social coverage" depicted in Figure 5.3.

If optimization has occurred at all, the relationships postulated to exist between physical specialization, behavioral flexibility, and social coverage should be demonstrable in nature. If the optimum approached is unconstrained in nature, that is if the species has responded to natural selection so as to match each task with a caste, then the number of physical and temporal castes should equal the number of tasks, and when many species are compared, the number of physical castes should increase linearly with the number of tasks: $\mathcal{N}_C \sim \mathcal{N}_T$. On the other hand, if $\mathcal{N}_C$ has been selected for in the manner suggested by the allometric space model, the number of physical castes should increase as the logarithm of the number of tasks: $\mathcal{N}_C \sim ln\mathcal{N}_T$. These distinctions can be tested by close behavioral comparisons of monomorphic or weakly polymorphic species with congeneric species that are strongly polymorphic. We shall consider the relation between $\mathcal{N}_C$ and $\mathcal{N}_T$ from a different viewpoint in Section 6.5.

6.2. FORCES AFFECTING CASTE PROLIFERATION

In theory at least, both ergonomic and nonergonomic constraints can operate to limit the number of castes in evolution. Nonergonomic constraints arise from the counteraction of colony-level selection by individual-level selection. When in-

sects are committed to an extreme laboring caste, particularly
when their ovaries are eliminated to increase efficiency or their
labor must be performed away from the brood chambers, they
surrender their reproductive potential. Under a wide range of
conditions such caste formation will be resisted by individual-
level selection, even if it enhances performance of the colony
as a whole (see Chapter Three).

In addition there exists a set of biological properties that can
inhibit the proliferation of castes through their influence on the
functioning of the colony as a whole. We shall refer to these as
ergonomic constraints, because they represent mechanisms that
contribute to the efficiency of the colony. Three (holometabo-
lism, allometry, fidelity costs) are derived from the developmental biology of the colony members; four others (environmental
dispersion, task overlap, behavioral plasticity, ergonomic costs)
are ecological in origin.

(1) *Holometabolism*

In the social Hymenoptera final adult size is determined at
the end of larval growth. In some species it is at least approximately fixed at decision points that occur earlier (see Chapter
Four). As a result, no further adjustments in the size-frequency
distributions can be made through later molting. Such adjustments can be made among the working immature stages of
termites, but the capacity of even these insects is limited by the
existence of terminal adult instars. In either case there exists a
developmental lag, and colony requirements must be antici-
pated ahead of the final, adult molt. The greater the magnitude
and unpredictability of environmental fluctuations, the less
precisely can the colony anticipate its needs and the fewer the
specialized physical castes that will comprise the optimum mix.
In the language of inventory theory, the "order lag time," the
period between the caste-determining decision point and adult
eclosion, must be shorter than the period of environmental
fluctuation, or else the colony optimum will be monomorphism.

(2) *Allometry*

It is a remarkable fact that the anatomy of the physical subcastes of each ant species is determined by a single rule of topological deformation. There is only one allometric series, so that the colony is limited to a single series of medias and a single form of major. Had multiple allometric series been evolved, far richer arrays of castes would have been easy to manufacture. As discussed in Chapter Four, the size-frequency distributions of adult workers within colonies are equally simple in quality. They can be explained as the outcome of elementary rules of relative growth with the occasional addition of one or two decision points during larval development. Why have relative growth and size-frequency distributions of ants remained this elementary? There may exist certain canonical constraints in the developmental biology of holometabolous insects, such that the growth transformation rules can be made no more complex. But it is at least equally possible that elementary allometric deformations and simple size-frequency distributions are all that are needed to produce the colony optimum mix. According to this second hypothesis, simplicity is imposed by constraints other than those intrinsic to developmental biology and is therefore a cause and not an effect.

(3) *Fidelity Costs During Development*

Individual workers are neither genetically identical nor subject to identical life histories during their development. In order to regulate the growth of immature forms with enough precision to achieve a particular uniform caste allometry, elaborate and sensitive feedback mechanisms would be required that correct continuously for the dispersion in growth of each age cohort. For example, larval feeding by nurse workers tends to be a haphazard process, so that an individual's nutritional history—and therefore its growth rate—is subject to random variations. This effect alone would create an adult size variation even if all other factors were held constant. The dispersionary influence can be entirely counteracted only by a much more sophisticated mode of brood rearing and concomitant behavioral and physiological mechanisms which, in turn, would require a great deal of genetic programming to implement. The "cost" of such mechanisms is difficult to assess, but might be measured in information-entropy units. It seems improbable to us that natural selection has led to such a precise control system. A redundant, "sloppy" mechanism is an equally reliable solution, and in any case the ergonomic return on such precision is likely to be negligible.

(4) *Environmental Dispersion*

The size requirements for a particular task are not monomorphic. Even in a relatively homogeneous, predictable environment, food particles and enemies vary continuously in size and other qualities in ways that affect their resistance to attack. Under many circumstances it will be advantageous to the colony to possess a built-in size variance in its worker caste that matches the predictable variability in food resources and enemies. To the extent that such anatomical variation is increased, the number of discrete castes that can be maintained will be diminished.

(5) *Task Overlap*

Some otherwise very disparate tasks probably require anatomical features that are closely similar. For example, if the colony's own pupae are approximately the same size and conformation as the prey on which the species feeds, pupal transport and prey retrieval will have similar optimum arrays of workers. With developmental errors reducing the differences still more, the optimal solution for the species might easily be a single array for both tasks.

191

(6) *Behavioral Plasticity*

Individuals of a given size can and are frequently required to perform tasks for which they are not specialists. When a nest of *Pheidole dentata* is broken open, a large fraction of the worker force responds by hurriedly carrying the brood to darker, safer retreats. At this time a few of the major workers join, even though they never attend brood under ordinary circumstances. In most insect colonies a large minority of the workers are either idle or else slowly patrolling the interior of the nest at any given moment. These have been interpreted as "reserve troops" that can be called on for general emergencies such as damage to the nest, invasion by enemies, or severe overheating in the brood chambers (Lindauer, 1961; Michener, 1964). As a consequence, the optimal mix must contain generalists as well as specialists, and the pressures for extreme discretization and proliferation of castes are correspondingly weakened.

(7) *Ergonomic Costs*

(a) The number of castes may be kept well below the number of required social acts as a concession to simple spatial efficiency. Consider a caste specialized to care for eggs. Eggs are scattered around the queen who lays them, and around the pupae, as well as around the new adults ecloding from the pupae. The mean free path between the objects attended by a worker will be smaller if the worker cares only for the queen, pupae, and ecloding nestmates together; this effort will be more economically allocated since the transition times between tasks are reduced. Minor workers of *Pheidole dentata* have in fact arrived at the latter solution. They pass through three temporal stages in which specialization occurs on those sets of tasks that are spatially the most nearly contiguous. Consequently the number of temporal castes are three rather than the twenty or more possible.

(b) Another kind of ergonomic constraint is imposed by the energetic cost of producing physically large castes. Physical castes in ants are based for the most part on allometry, which depends in turn on an overall size increase. With each increment in a linear dimension, such as pronotal width or antennal length, there is approximately a cubed increase in weight—and therefore in the energetic cost of manufacture. Remarkably, the maintenance (i.e., metabolic) cost does not necessarily go up to a corresponding degree. Major workers of the Florida harvester ant *Pogonomyrmex badius* have about the same resting rate of oxygen consumption as minor workers, despite the fact that they individually weigh ten times as much. Their consumption per gram is simply one-tenth that of the smaller caste (Golley and Gentry, 1964). *Myrmica rubra* queens weigh 2.3 × as much as workers but consume only 1.8 × as much sucrose (Brian, 1973). Nevertheless, the combined manufacture and maintenance costs do go up disproportionately with an increase in linear dimensions and must eventually produce a constraint on the number of new castes. Some ant species employ an anatomical trick to reduce the burden: an exceptionally strong allometry yields major workers with thoraces and abdomens of only slightly increased size surmounted by grotesquely large heads and mandibles, the body parts most directly concerned with caste specialization. The ultimate example is the major worker of *Acanthomyrmex*, whose head overhangs the thorax to the rear and weighs approximately as much as the remainder of the body (see Figure 6.4). But this economy only emphasizes the problems involved in a physical system based on size and allometry, and it suggests that caste proliferation is indeed occasionally limited by metabolic cost.

6.3. THE CORRELATES OF CASTE PROLIFERATION

The forces inhibiting the proliferation of castes are so powerful that the workers of most species of ants, as well as those of virtually all social bees and wasps, are held at anatomical

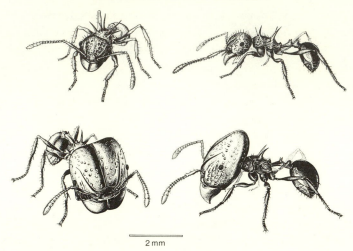

FIGURE 6.4. Minor and major workers of a Celebes species of *Acanthomyrmex*. (Drawing by Turid Hölldobler.)

monomorphism combined with a limited differentiation into temporal castes. Now we ask, what are the countervailing forces that move species toward a polymorphic equilibrium?

A partial answer can be supplied immediately with respect to physical castes. The shearing forces appear to be relatively few and easy to characterize. They are so distinctive as to appear special and powerful; in other words, the conservative constraints leading back to monomorphism have not been easy to overcome. The following generalizations apply to ants; termite biology is still too imperfectly known to summarize with the same degree of confidence.

(1) *Majors Are the Novel Caste*

Physical polymorphism has usually been created by the addition of a major caste, based on an allometric enlargement and reshaping of the head. A few genera, including *Atta*, *Eciton*, and *Pheidologeton*, have added a minim caste at the other extreme of the size range. *Oecophylla* has added a minim caste to the more

"normal" regular worker caste without the simultaneous creation of a major caste.

(2) *Majors Have Limited Roles*

Major workers of various ant species appear to arise only as
specialists for one or the other of four primary tasks: foraging,
defense, food storage (by the excessive distension of the crop to
create repletes), and the milling (chewing) of seeds.

(3) *Extreme Majors Are Most Specialized*

Anatomically very deviant major forms appear to be specialized either for defense or for milling, to the exclusion of
other roles. On the other hand, there are a few species in which
the major caste functions in defense or milling but is not very
deviant.

(4) *Milling Does Not Always Require Majors*

It is possible that an extreme miller caste arises in ant species
only when seeds constitute a substantial supplement to the diet
yet are not the principal or sole dietary component. This
particular state is known to be the case in *Solenopsis geminata* and
is likely or probable in some species of *Acanthomyrmex* (W. L.
Brown, in preparation) and *Pogonomyrmex badius* and some
species of *Pheidole*. Where seeds are the principal or sole dietary
items, it appears that the species is monomorphic or weakly
polymorphic. In other words, all of the workers are adapted to
milling, yet remain generalists in other tasks and hence more
"normal" in appearance. Examples of the second category are
believed to include the species of *Veromessor* and *Monomorium*
subgenus *Holcomyrmex*.

(5) *Minim Workers Are Nurses*

When a distinct minim caste persists after the earliest stages
of colony growth, it is specialized for care of the brood, particularly of the eggs and smaller larvae. In leafcutter ants of the

genus *Atta*, minims also serve as fungus gardeners, visiting cavities in the fungus beds too small to admit the larger castes.

By comparing many species of ants, the following qualities are found to be correlated: the size of the mature colony, the number and complexity of tasks performed by the worker caste, the size range of the worker caste within colonies, and the number of recognizable physical subcastes. In short, the more complicated the mode of life, the greater the physical polymorphism. This relation cannot be measured yet in a quantitative way, but it is clear enough when reviewing the natural history of the best known ant species. The following three species can be taken as representatives of simple, moderate, and complex "life styles" respectively.

Strumigenys louisianae. These small ants can be taken as representative of a large class of ponerines and myrmicines that form mature colonies of only one to several hundred workers and prey on other arthropods (in this case, mostly on collembolans). Foraging is conducted entirely by single workers; chemical trail systems and other forms of recruitment are unknown. The nests consist of several irregular chambers and galleries in rotting pieces of wood and the soil. The queen and relatively small brood populations are found closely grouped together. There are approximately twenty distinguishable tasks, all of which are performed by a monomorphic worker caste (see Brown and Wilson, 1959).

Pheidole dentata. The mature colonies contain between 2,000 and 4,000 workers, which are common values for the large number of ant species occupying the middle range of population size. The workers capture arthropod prey belonging to a diversity of taxonomic groups and scavenge dead arthropods of all but the smallest sizes. They also collect sugary materials encountered on their foraging trips but are not known to attend honeydew-producing insects. Well-formed nests are excavated in pieces of rotting wood, with chambers and galleries occasionally extending into the soil. The brood is segregated into two groups according to its stages of development, and the groups are attended by workers of different ages. Twenty-six tasks have been identified. There is a distinct major worker subcaste, which is recruited by minor workers to repel invaders and to protect newly discovered large food sources outside the nest. (See Wilson, 1975c, 1976a.)

Atta laevigata. This is a typical species of leafcutter ants in the genus *Atta.* Colonies of these obligate fungus growers are complex agricultural states. At maturity they contain from a hundred thousand to over one million adults. Foragers travel daily over networks of chemical trails to gather fresh leaf fragments, which are chewed, manured, and incorporated into the substratum of the fungus gardens. The diet of the ants consists mostly or entirely of a single species of symbiotic fungus cultivated by this method. The nests, which penetrate as much as six meters into the soil, consist of several kinds of chambers, each with a separate function, together with interconnecting galleries. The nest microclimate is more evenly regulated than in the case of *Strumigenys* and *Pheidole.* The number of tasks probably exceeds fifty. The maximum size range of the worker caste is the greatest found in the ants. The size variation is continuous and encompasses morphologically very distinct minima, media, and major subcastes (see Figure 6.5). The number of true physical subcastes definable along the size gradient, according to the criteria proposed earlier, may prove to be still greater. (See Weber, 1972, for a general review of the biology of *Atta.*)

The problem of caste multiplication within a colony has certain parallels with the problem of species packing in faunas and floras. If a community of species were to live in a strictly unvarying environment, it would be theoretically possible for additional species to be inserted until each species contained the least number of individuals that meaningfully constitute a population. The species would merely divide up the available

197

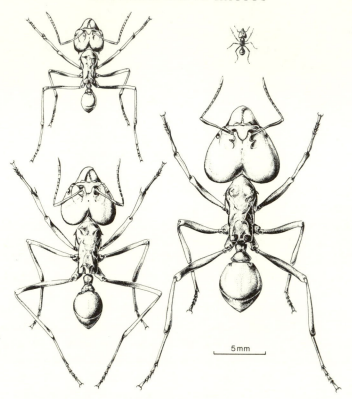

FIGURE 6.5. The caste systems of leafcutter ants in the genus *Atta* are among the most complex in the social insects, involving extreme size variation accompanied by strong diphasic allometry. The species illustrated here is *A. laevigata*. (Original figure by Turid Hölldobler.)

resources, for example food particles of varying size, to an ever finer degree. But in the real world, where the environment fluctuates, a much lower limit on species numbers is set. Here, in general, the maximum number is reached when the average food particle sizes for species adjacent on the resource spectrum differ by an amount approximately equal to the standard deviation in the food size taken by either of the species (May, 1973).

198

The "caste niche" within colonies of social insects is also theoretically very finely divisible. In a deterministic environment a perfect caste can be designed for each task, right down to a subdivision of foraging or nest construction so as to specialize on particles of different sizes. In the extreme conceivable case, each member of the colony would constitute a separate caste. But constraints do exist, and they hold the actual number of castes far below even the broad categories of tasks confronting the colony. To a large extent these forces stem from stochastic fluctuations in the environment, but they are more numerous and diverse than those controlling species diversity within ecosystems. At present there is no precise way to assess their relative importance. On a largely intuitive basis we suggest the following rank order, from the most to the least potent in caste evolution:

I. ECOLOGICAL CONSTRAINTS (ERGONOMIC)
    1. Ergonomic costs
    2. Environmental dispersion
    3. Behavioral plasticity
    4. Task overlap
II. OPPOSING LEVELS OF NATURAL SELECTION
    (NONERGONOMIC)
    5. Individual-level selection
III. DEVELOPMENTAL CONSTRAINTS (ERGONOMIC)
    6. Allometry
    7. Fidelity costs in development
    8. Holometabolism

In making the above conjecture we have assigned a lower weight to the developmental constraints because of the extreme anatomical diversity that occurs among closely related species of insects and between the sexes of individual species. These are frequently associated with narrow anatomical and behavioral variances. Also, the zooids of colonies of ectoprocts, siphonophores, and other colonial invertebrates are very diversified.

199

Since they are also genetically identical within colonies, they represent the potential of caste evolution with the constraint of individual-level selection removed. Individual-level selection is in turn rated below the direct environmental constraints because of the evidence that it can be removed without an explosion of caste proliferation. In the genera *Monomorium*, *Pheidole*, and *Solenopsis*, workers lack ovaries and hence entirely surrender their personal reproductive potential. These species are more frequently polymorphic, but the degree of their polymorphism does not exceed that of many other species whose workers possess well-developed ovaries.

6.4. OPTIMUM SIZE-FREQUENCY DISTRIBUTIONS

When the sociograms of many ant species are completed, the general relation between modes of life and the complexity of caste systems should become much clearer. The pressing need now is for careful empirical analysis. Therefore, rather than attempting to formulate global models of caste structure—which cannot be empirically validated in the foreseeable future—we shall limit ourselves to reduced models that (a) address particular questions, (b) are defined in operational terms so as to provide a conceptual framework for experimental work, and (c) are amenable to empirical disproof.

We shall first construct a very simplified model that deals only with the marginal size distribution, $n(s)$. When the appropriate empirical observations become available, more elaborate calculations on the complete CDF can be carried through that follow the same pattern of analysis.

The size distributions of all eusocial insects fall roughly into one of three categories: (1) unimodal, often with skewing toward the larger size classes, (2) bimodal (see for example, Figures 4.3 and 4.4), and (3) trimodal. Type (1) is by far the most common while type (3) has been observed in only several species. We shall employ the following model suggesting that this pattern may be the result of purely ergonomic forces.

200

Consider a colony of fixed age, $\overline{t}$, and average overall age cohorts and tempo:

$$n(\mathbf{s}) = \frac{1}{\mathcal{N}} \iint n(\overline{t},a,\mathbf{s},x)\ da\ dx.$$

Imagine this colony immersed in an environment which supplies resources at a constant rate with size distribution $R(\hat{s})$, and which must be harvested by the colony's foraging force. For the moment we neglect all contingencies except foraging.

In Figure 6.6 we show an imaginary empirical resource size distribution of arthropod prey items, $R(\hat{s})$. Denote by $\eta(s,\hat{s})$ the foraging rate of an individual of size s on particles of size $\hat{s}$. One expects $\eta(s,\hat{s})$ to be distributed about s in the fashion shown in Figure 6.6 (e.g., a bell-shaped curve). The total profit rate from foragers of size s working the resource spectrum, $R(\hat{s})$ is:

$$P(s) = \int_0^{\infty} R(\hat{s})\eta(s,\hat{s})\ d\hat{s}. \tag{6.7}$$

The total foraging return is then given by

$$\text{foraging return} = \int_0^{S_m} P(s)n(s)\ ds, \tag{6.8}$$

where S_m is the maximum size of workers.

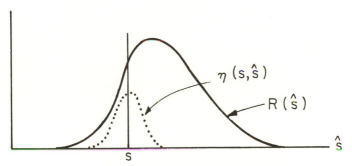

FIGURE 6.6. A hypothetical size distribution of food particles, $R(\hat{s})$, and foraging efficiency, $\eta(s,\hat{s})$. The foraging profit from a worker of size s is given by $\int R(\hat{s})\eta(s,\hat{s})\ d\hat{s}$, where η is the foraging efficiency function (FEF) and is shown as the dotted line.

From this must be subtracted the maintenance, manufacturing, and fidelity costs discussed earlier. The manufacturing costs $C_m(s)$ are expected to rise as the mass: $C_m(s) \sim s^3$. The maintenance costs, surprisingly, remain fairly constant with size in at least one species, *Pogonomyrmex badius* (Golley and Gentry, 1964). This may be due to the fact that larger castes tend to be more sedentary, rousing themselves to rapid activity when recruited by minor workers. (In Chapter Seven we will give an example of how a foraging efficiency function can be derived from a time and energy budget. The technique employed includes the metabolic costs, in the expression for η.) The fidelity costs are more difficult to assess. In principle, they can be estimated from measurements of growth dispersion in a constant environment. For our purposes we will model the fidelity costs by an expression of the form

$$\text{fidelity costs} = k \left(\frac{dn}{ds}\right)^2. \tag{6.9}$$

That is, we impose a penalty against abrupt changes in the caste profile. The constant k measures the cost associated with creating a sharply delineated size class. Thus, the optimum caste distribution $n^*(s)$ is determined by solving the following programming problem

$$\underset{n(s)}{\text{Max}} \int_0^{S_m} [(P(s) - C_m(s))n(s) - kn'(s)^2]\, ds, \tag{6.10a}$$

subject to the constraints

$$\int [C_m(s)n(s) + kn'(s)^2]\, ds \leqslant M \tag{6.10b}$$

$$n(s) > 0, \qquad 0 < s < S_m.$$

Here m is the total available resources to be allocated among the various foraging size classes. In Appendix 6.3 we show that the optimum size distribution is given by

$$n^*(s) = \frac{1}{2k(1-a)} \int_0^s \int_s^{S_m} [P(t) + aC_m(t)]\, dt\, ds, \tag{6.11}$$

202

where the constant $a < 0$ is determined from the normalization condition (6.10b).

In Figure 6.7 we have plotted $n^*(s)$ for a hypothetical resource distribution and the efficiency function computed in Chapter Seven. Clearly, if no costs were imposed, the ideal caste distribution would be concentrated in a "spike" at the peak of $R(s)$, since the colony can do no better than to harvest the most remunerative resource size class. The extent to which the caste distribution $n^*(s)$ deviates from $R(s)$ is a measure of: (1) the countervailing costs, (2) the strength of conflicting contingencies, (3) the magnitude of environmental uncertainties, and (4) the dispersion in caste efficiencies (i.e., behavioral

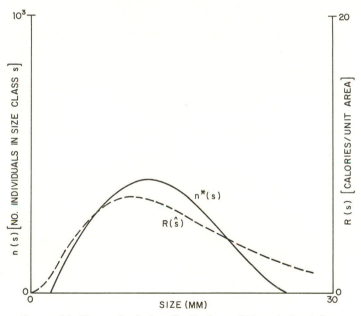

FIGURE 6.7. The optimal size distribution, $n^*(s)$, calculated from Equation (6.11), corresponding to a resource distribution, $R(\hat{s})$, as shown. Note that depending on the shape of the foraging efficiency function, $\eta(s,\hat{s})$, there need be no simple relationship between $R(\hat{s})$ and $n^*(s)$.

flexibility). This last effect is, we feel, probably the dominant selective force preventing a stable and optimum match between caste sizes and resource characteristics. In the following sections we will address this issue explicitly and develop some conceptual tools for assessing its significance. For the purposes of our present discussion, however, a static, deterministic model may be sufficient.

Using the model it is easy to see why unimodal size distributions are by far the most common type encountered in nature. The manufacturing costs rise quite fast with size, and so, unless a particular resource is extraordinarily remunerative, it will not pay to produce a matching size distribution of workers. The situation is represented schematically in Figure 6.8: if a second resource, $R_2(\hat{s})$, is available to a species (e.g., seeds versus nonseed items, cf. section 6.4) then the CDF will be bimodal only if the profit exceeds the costs. Moreover, the second "hump" in the CDF corresponding to the larger size classes will almost always be much smaller than the CDF hump corresponding to the primary RDF.

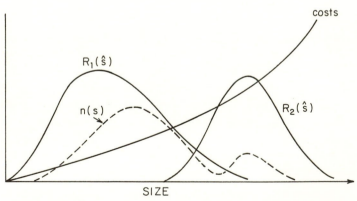

FIGURE 6.8. Due to the rapidity with which costs increase with size, multimodal size distributions rarely result from multimodal resource distributions.

Thus, the optimum resource matching model suggests two general conclusions concerning the CDF.

(a) The shape of the optimal caste size distribution cannot be directly inferred from the resource distribution. The various costs enter in a nonlinear way and distort the CDF such that the first several moments (mean, variance, skewness, kurtosis, etc.) of $n(\hat{s})$ are usually different than those of $R(\hat{s})$. In order to test the adaptive optimality of an observed CDF one cannot simply compare it with the observed RDF; rather, the optimization procedure outlined in Appendix 6.1 must be carried out on the RDF, and the calculated optimum CDF compared with the observed CDF.

(b) Because the costs rise so rapidly with size, one expects unimodal CDF's to be the most common type. In those cases where bimodal or trimodal size distributions are observed the larger size classes are invariably either (i) associated with processing an extraordinarily rich resource, such as the seed-grinding major caste in *Solenopsis geminata* (cf., section 6.3); or (ii) they form a soldier caste specialized to contend with unavoidable defensive contingencies.

6.5. THE EFFECT OF UNCERTAIN ENVIRONMENTS ON CASTE SPECIALIZATION

In ant colonies belonging to the genera *Acanthomyrmex*, *Pheidole*, and *Solenopsis*, the major workers are considerably larger than minors and their mandibles are specialized for the milling of seeds. They are also behaviorally specialized; in the case of the fire ant *Solenopsis geminata* they respond only to the presence of seeds and leave to the minor caste the task of collecting and processing all other kinds of food. One can readily predict that species adapted to an environment where seeds are the dominant food source will tend to evolve major castes, while those adapted to seed-poor environments will be monomorphic. However, as mentioned in section 6.2, beyond a certain seed abundance the degree of caste specialization may actually decrease. In order to understand how such a paradoxical reversal is possible, consider the simple model represented in Figure 6.9.

First we discretize the size distribution as shown in Figure 6.9a to consider just two castes: majors (M) and minors (m). Majors are assumed to process only seeds, while minors can process seeds and other food items as well. Denote by η_m the utilization efficiency of minors on non-seed items and by η_M the efficiency of majors on seeds. Minors utilize seeds with an efficiency $k\eta_M$, where $k < 1$. Thus, k measures the decrease in efficiency concomitant with being a generalist. The environment supplies seeds and other food items at a steady stochastic rate $S(t)$ and

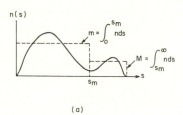

(a)

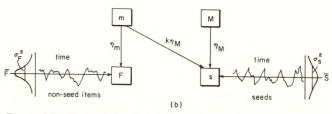

(b)

FIGURE 6.9. The analysis of the effects of an uncertain food supply begins with the discretization of the worker population into two castes, minors m and majors M, as depicted in the upper diagram (a). In the lower diagram (b) the food is classified into two categories, seeds S and items of other kinds F; the supply of each category through time is represented schematically. Majors process seeds exclusively, with an efficiency of η_M. For the most part minors utilize other kinds of food η_m but they are also able to handle seeds to some extent (that is: $k\eta_M$ where $k < 1$).

$F(t)$. We shall assume that the stochastic processes $F(t)$ and $S(t)$ can be characterized by the means and variances of their associated probability distributions, which are designated

$$\{\overline{F}, \sigma_F^2\} \text{ and } \{\overline{S}, \sigma_S^2\}.$$

Since it is likely that these processes are correlated, σ_{FS}^2 denotes the covariance between the seed and food supplies. Then the net energetic profit rate from foraging can be written

$$\mathcal{J} = [\text{Foraging profit}] - [\text{maintenance}] - [\text{manufacturing}]$$

$$= [m(\eta_m F + k\eta_M S) + M(\eta_M S)]$$

$$- [r_m m + r_M M] - [c_m m + c_M M]$$

where (r_m, c_m) and (r_M, c_M) are the metabolic and manufacturing costs of minors and majors, respectively.

As shown in Chapter Five the colony fitness, which is measured by the number of reproductive alates (Q) produced by the time of the nuptial flights, is an increasing function of the energetic resources accumulated during the ergonomic phase. The optimum caste ratio is obtained by maximizing the expected fitness, $\langle Q \rangle$, subject to an ergonomic constraint on the total amount of resources and labor available. Our purpose, then, is to examine the following simple optimization problem in the presence of uncertainty:

$$\text{Max} \langle Q(\mathcal{J}) \rangle,$$

subject to

$$M + m \leqslant \mathcal{N}$$

$$M, m \geqslant 0,$$

where $\mathcal{N}$ is the maximum total foraging and food-processing force, and where $Q(\cdot)$ is a monotonically increasing function of the net foraging profit rate.

Since F and S are random variables, so are $\mathcal{J}$ and Q. The following approximate analysis will illustrate the points we wish to emphasize. First, we expand Q about $\langle \mathcal{J} \rangle$ to second

207

order and take expectations of both sides to obtain:

$$\langle Q(\mathcal{J}) \rangle = Q(\langle \mathcal{J} \rangle) + \rho\sigma^2 + \cdots$$

where σ^2 is the variance in $\mathcal{J}$ and $\rho = (1/2)Q''(\langle \mathcal{J} \rangle)$ will be called the *risk factor*. (Note that when ρ is positive, the $Q(\mathcal{J})$ curve is turning upward. Consequently, a positive payoff in $Q(\mathcal{J})$ yields more than a negative one loses; the average payoff is positive, and the gamble worth taking. When ρ is negative, the reverse situation exists; see also Chapter One.) If the shape of $Q(\cdot)$ can be approximated by a saturating function of the form

$$Q(\mathcal{J}) = Q_0(1 - e^{-\rho J})$$

and the random variables F and S are Gaussian, then it can be shown that maximizing $Q(\langle \mathcal{J} \rangle) + \rho\sigma^2$ is equivalent to maximizing $\langle Q(\mathcal{J}) \rangle$, and so the 2-term expansion for Q is exact.[1]

From the expression for the foraging rate we can write for the mean and variance of $\mathcal{J}$.

$$\langle \mathcal{J} \rangle = (m,M)\left\{ \begin{bmatrix} \eta_m & 0 \\ k\eta_m & \eta_M \end{bmatrix}\begin{bmatrix} \bar{F} \\ \bar{S} \end{bmatrix} - \begin{bmatrix} r_m + c_m \\ r_M + c_M \end{bmatrix}\right\} \triangleq \mathbf{x}^{\mathrm{T}}\mathbf{A}$$

$$\sigma^2 = (m,M)\begin{bmatrix} \sigma_F^2 & \sigma_{FS}^2 \\ \sigma_{FS}^2 & \sigma_S^2 \end{bmatrix}\begin{bmatrix} m \\ M \end{bmatrix} \triangleq \mathbf{x}^{\mathrm{T}}\sigma^2\mathbf{x},$$

where $[\sigma^2]_{ij}$ is the covariance matrix between the random processes F and S.

Now, for the moment let us examine the consequences of maximizing only the mean foraging return, $\langle \mathcal{J} \rangle$; that is, we solve the programming problem

$$\mathrm{Max}\langle \mathcal{J} \rangle$$

subject to

$$M + m \leqslant \mathcal{N}, \qquad m,m \geqslant 0.$$

This is illustrated in Figure 6.10a. When $\mathbf{1}^{\mathrm{T}}\mathbf{A} < 0$, i.e., $\eta_m S - r_m - c_m < \eta F + k\eta_m S - r_m - c_m$ the mean net profit is maximized by producing only minors; this is the case illustrated in

208

Figure 6.10a. Conversely, when the average net profit derived from seeds exceeds the return for food other than seeds, then the optimum strategy is to specialize in majors. This is an obvious consequence of our assumption that foraging efficiencies are constant: the maximum average return is obtained by specializing on the caste with the largest average productivity. However, this strategy neglects the role of environmental unpredictability. The decision is analogous to that made by a storekeeper who only stocks to meet the average demand; because of the variability in demand, half the time he is caught out of stock. Thus, ignoring the second term in the expansion of $\langle Q \rangle$ is tantamount to assuming a perfectly predictable environment. The variances σ_{ij}^2 are reasonable measures of environmental uncertainty, or "risk." In Figure 6.10b we have superimposed a family of constant risk contours onto the $(M\text{-}m)$ plane. Since σ^2 is a quadratic form, the contours are a family of ellipses centered at the origin.[2] From this Figure it is easy to see that the maximum return, at S_1, corresponds to the maximum risk ellipse (consistent with the constraint $m + M < \mathcal{N}$), while decreasing risk can only be obtained at the cost of decreasing returns. Note that the minimum risk ellipse touches the constraint at g, which is higher than the minimum return located at s_2. Therefore, when $\mathbf{1}^T\mathbf{A} < 0$, caste ratios on the line $(g\text{-}s_2)$ are never preferred; one can always obtain a higher return at the same risk level. Optimized caste ratios should lie only on the $(s_1\text{-}g)$ segment of the constraint.

Along $(s_1\text{-}g)$ there is a continuous tradeoff between risk and return. This tradeoff is shown explicitly in Figure 6.10c, where the segment $(s_2\text{-}g\text{-}s_1)$ is plotted in $(\langle \mathcal{J} \rangle, \sigma^2)$ coordinates. It is easy to show that the fitness set $(s_1\text{-}g)$ is convex as shown, i.e.,

$$\frac{d^2\sigma^2}{d\langle \mathcal{J} \rangle^2} > 0$$

(cf., Markowitz, 1959; Mossin, 1968). *An optimum caste ratio cannot be selected until a risk level, $\rho\sigma^2$, has been selected.* For a given

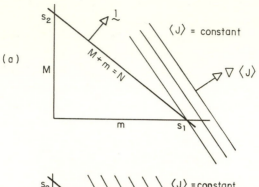

(a)

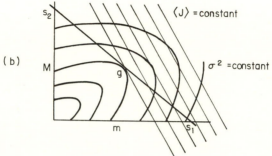

(b)

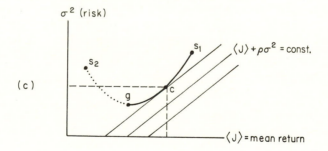

(c)

FIGURE 6.10. The foraging profit $\mathcal{J}$ depends on the investment in castes specialized to exploit various categories of food sources (refer to Figure 6.9). (a) If the flow of the food sources is constant so that only the mean foraging profit matters, the optimum solution is total investment in one caste alone. In other words, the maximum profit contour, $\mathcal{J} = $ constant, occurs at the monomorphic point s_1 on the constraint $M + m = \mathcal{N}$. (b) If the food supply is variable, the variance of the income σ^2 and hence the risk increases as heavy investment is made in one caste at the expense of the other. The risk curves form a series of outward expanding ellipses. Consequently the species must "choose" a compromise apportionment of castes that lies somewhere on the segment running from low risk, low average yield g to high risk, high average yield s_1. Only caste compositions between g and s_1 are preferred; caste compositions between g and s_2 have lower yields for the same risk level. Along g-s_1 increasing return must be paid for by higher risk levels. (c) By plotting the tradeoff curve in $\langle \mathcal{J} \rangle$, σ^2 coordinates, the tradeoff between risk and return can be shown explicitly. The optimum caste composition, C, is determined once an overall fitness function is selected. Here we have shown the fitness contours corresponding to a linear weighting of mean return, $\langle \mathcal{J} \rangle$, and risk, σ^2, with a risk coefficient of ρ.

ρ, the caste ratio is determined by maximizing $\langle \mathcal{J} \rangle - \rho \sigma^2$ along the constraint $m + M = \mathcal{N}$. The constrained optimum is given by

$$0 < \frac{m}{M + m} = \frac{\bar{F} - \bar{S} + 2\rho(\sigma_{FS}^2 - \sigma_S^2)}{2\rho(2\sigma_{FS}^2 - \sigma_F^2 - \sigma_S^2)} < 1.$$

The model also predicts that *tychophobe species*[3]—those with very concave fitness and hence least likely to take risks—will have caste ratios stable about the polymorphic point g. However, consider a *tychophile species* characterized by higher risk acceptance, as shown in Figure 6.11. As the resource composition is varied, the slope of the mean return lines changes. At a critical value of the resource composition the optimum caste ratio jumps from point a to point b. This suggests that species adapted to higher risk levels (or more uncertain environments)

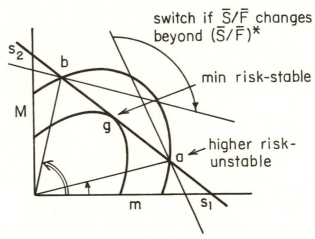

FIGURE 6.11. A "tychophobe" species that has chosen low risk at the price of a reduced potential yield (point g) is likely to have a more stable polymorphism in evolution. A "tychophile" species, one that chooses a higher risk strategy (point a), is more likely to switch to a new position (point b) in response to a change in the proportions of available resources, in this case denoted by the mean values of two kinds of food, $\bar{S}$ and $\bar{F}$. In high-risk species the colonies may also show greater flexibility in caste composition subject to environmental influence.

should vary among themselves more in caste compositions—or even that the colonies of such species should have less canalized caste compositions. Moreover, shifts to dramatically different caste ratios as the frequency of environmental contingencies change should be observed among tychophile species.

Thus, the model not only suggests a rationale for the empirical observation of a limit on polymorphism but, more importantly, it provides a scheme for classifying certain aspects of a colony's adaptive strategies. Specifically, we can linearly order the life styles of species along the risk-return frontier according to how they are adapted to cope with environmental uncertainty. The tychophobe-tychophile continuum is somewhat analogous to the r-strategist versus K-strategist classification derived from classical Volterra-Lotka theory (see Figure 6.12).

Recall the basic argument on system reliability given in Chapter One. There it was shown that a convex fitness function characteristic of beginning colonies implies a selective force for risk taking, since the expected fitness resulting from a "gamble" exceeded the actual fitness if the gamble were declined. Beginning colonies are on the lower limb of the sigmoid growth curve. A "gambling loss" of $-\Delta N$ in colony size reduces the rate of colony growth by a relatively small amount, while a comparable gain of $+\Delta N$ results in a substantial increase in the growth rate, since it moves the colony up the steeply ascending limb of the colony growth curve. Conversely, concave fitness functions favor risk aversion, while linear fitnesses are risk-neutral (Figure 6.12a).

Thus, tychophilic and tychophobic social systems could be employed at different stages of the life cycle of the same colony. In addition, we might expect to find entire species committed primarily to one or the other of the two extreme strategies, as suggested in Figure 6.12b. Species whose colonies are conspicuously both fast-growing and short-lived, such as the wasps *Mischocyttarus* and *Polistes*, are more likely to have tychophilic caste proportions and behaviors. Tychophobic adaptations are

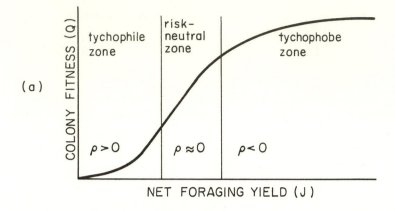

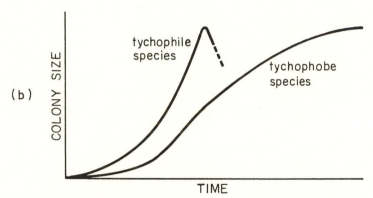

FIGURE 6.12. The suggested biological correlates of tychophile (risk-taking) and tychophobe (risk-avoiding) social strategies. (a) When the risk adaptedness factor $\rho[\triangleq 1/2Q''(\langle \mathcal{J} \rangle)]$ is positive, it is advantageous to try to take advantages of resource variation in order to maximize $\mathcal{J}$, leading to tychophile behaviors and caste systems; when ρ is negative, a tychophobe strategy is preferred. (b) Tychophily is expected in species with rapid colonial growth and a relative short average life span of colonies, while tychophobe strategies should occur more frequently where slow-growing, long-lived colonies are the rule.

best sought in the species whose colony growth rate begins to decline relatively early in the colony life cycle, such as the leafcutter ants of the genus *Atta*, as well as *Nasutitermes* and certain of the other higher termites.

Using the same model we can now assess the general principle of growth strategy with reference to the optimization of caste ratios and thus characterize the optimum social strategies more explicitly. Recall that since $\langle \mathcal{J} \rangle$ is a monotonically increasing function of colony size, and since

$$\rho \triangleq \tfrac{1}{2} Q''(\langle \mathcal{J} \rangle),$$

we see that if $Q(\cdot)$ is convex, $\rho > 0$. In computing the optimum of $\langle Q \rangle$ the variance is weighted positively, producing caste ratios that lie on higher risk contours. Thus, convex fitness implies a high risk factor. If the environmental fluctuations affecting the two castes are reasonably well correlated, that is, if $\sigma_{SF}^2 \sim \sigma_S^2 + \sigma_F^2$ (the situation shown in Figure 6.11), then the "generalist" point g will be close to the major axis of the risk ellipses, corresponding to a caste ratio of

$$\frac{M}{m} \sim \frac{2\sigma_{FS}^2}{\sigma_F^2 + \sigma_S^2}.$$

As ρ increases the optimum caste ratio moves toward s_1. If g is above the 45° line the caste polymorphism will increase, and then decrease until, at very high risk values, the monomorphic point s_1 is reached.

It would be tempting to make the generalization that a high risk parameter (very convex fitness) implies monomorphic caste structures. However, as the model shows, the optimum caste ratios depend on the variance and covariance of all of the task contingencies. In general, the optimum caste proportions are found by solving the quadratic programming problem: $\text{Max}\langle \mathcal{J} \rangle + \rho\sigma^2$, subject to $M + m \leqslant \mathcal{N}$, $M, m \geqslant 0$, and where the risk parameter is obtained from the fitness $Q(\mathcal{J})$. In cases where the contingencies have comparable variances and are at least moderately correlated (see, for example, Figure 6.10), one would expect incipient colonies to "choose" caste monomorphism. Indeed, this appears to be the case for many species: caste differentiation only appears after the colony has become well established.

When more than two castes are present a convenient measure of the caste diversity of a colony is

$$D \equiv \prod_{i=1}^{C} (1 + f_i),$$

where $f_i = N_i/N$ (the fraction of individuals of caste i) and C = total number of castes. An invariant scale for caste diversity in a population is the "information" measure,

$$\langle \log D \rangle = \sum_{i=1}^{C} f_i \log (1 + f_i),$$

which generalizes for the case of continuous caste (size) distributions to $\int f(s) \log(1 + f(s)) \, ds$, where s is size. If the environmental variances are known it might be possible to plot caste diversity, D, as a function of risk adaptedness, ρ. For the case given in Figure 6.10 this will resemble the plot shown in Figure 6.13a. Conversely, the model predicts that for fixed ρ (e.g., a given colony size), a comparative study of colonies in different environments will reveal caste diversity rising with σ^2. That is, increasing environmental uncertainty requires greater caste polymorphism in order for the colony to achieve homeostasis at a fixed risk level (Figure 6.13b). However, this expected

FIGURE 6.13. The correlation of caste diversity with risk acceptance and environmental uncertainty in food items.

correlation between environmental variability and caste diversity applies only to tasks with comparable variances.

In the case of general fitness functions, $Q(\mathcal{J})$, and non-Gaussian environments, the two-term expansion will not be sufficient to characterize the selective forces acting on caste proliferation. Referring again to Figure 6.10c, we see that optimizing $\langle \mathcal{J} \rangle + \rho \sigma^2$ is equivalent to choosing a linear utility function $Q(\langle \mathcal{J} \rangle, \sigma^2)$ shown in dotted lines. The slopes of these lines are determined by ρ, which, in turn, determines the optimal caste point, c. More general fitness functions can be devised, but in the absence of detailed empirical data this is probably not worthwhile pursuing further.

The mean-variance approach to modeling environmental uncertainty is attractive because of its simplicity and intuitive appeal. Nevertheless, it can be misleading in certain circumstances (Borch, 1968). For example, environmental fluctuations are seldom symmetric about their mean: food or attacks generally come in clumps in space and spurts in time. Thus, in expanding $Q(\mathcal{J})$, at least the third term should be retained to measure the "skewness" of the distribution. The usefulness of a specialized caste could be heavily influenced by the presence of occasional, but severe, contingencies.

6.6. SPECIALISTS AND GENERALISTS

The model developed in section 6.5 will now be used to address the more general question of how environmental variability affects the evolution of specialized castes.

Consider the following idealized system. A colony must perform different types of tasks, which are designated T_i $(i = 1, 2, \ldots, t)$. To cope with these requirements it can manufacture c distinct castes in amounts N_j $(j = 1, 2, \ldots, c)$. Denote by η_{ij} the efficiency of caste j in performing task i; the model is shown schematically in Figure 6.14 for the case of $t = 2$, $c = 3$. (Note that we are considering *possible* castes; thus, although

217

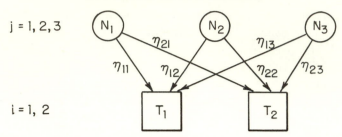

$j = 1, 2, 3$

$i = 1, 2$

FIGURE 6.14. The essential quantities of an elementary caste specialization model. In each colony, workers are allocated among the possible castes, in this case $j = 1, 2, 3$. These groups address two tasks, T_i, where $i = 1, 2$. Their efficiencies are designated η_{ij}.

provisional allowance is made for $\mathcal{N}_t < \mathcal{N}_c$ in this section, it will be shown that in fact $\mathcal{N}_t \geqslant \mathcal{N}_c$.) The rate at which task i is performed can be approximated by a "mass action" functional response model:

$$\mathcal{J}_i = \text{Rate of task } i = \sum_{j=1}^{c} \eta_{ij} T_i \mathcal{N}_j$$

The total performance rate of all tasks is then:

$$\mathcal{J} = \sum_{i=1}^{t} \mathcal{J}_i = \sum_{i=1}^{t} \sum_{j=1}^{c} \eta_{ij} T_i \mathcal{N}_j \triangleq \sum_{j=1}^{c} a_j \mathcal{N}_j = \mathbf{a}^T \mathbf{N}$$

where

$$a_j \triangleq \sum_{i=1}^{t} \eta_{ij} T_i = \text{total contribution of caste } j$$
$$\text{(measured, say, in cal.}/t\text{)}.$$

This is analogous to a discretized version of the foraging model presented in section 6.4.

Suppose that the colony has available to it a total of E calories in energetic resources and that the cost of manufacturing an individual of caste j is $c_j[\text{cal.}]$. Then the economic problem facing the colony is to determine the optimal allocation of its resources so as to maximize its fitness, $Q(\mathcal{J})$. That is, the species is expected to evolve toward

$$\text{Max } Q(\mathcal{J}),$$

subject to

$$\sum_{j=1}^{c} c_j \mathcal{N}_j \leqslant E, \qquad \mathcal{N} > 0.$$

If $Q(\mathcal{J})$ is convex, then the optimal solution will always be a "pure" strategy: make only one caste—that which yields the best performance/cost tradeoff. This is shown in Figure 6.15a. In a perfectly predictable environment this result is intuitively obvious: the colony can do no better than to invest in the caste that yields the highest net return.

However, in the context of the model several factors can act to promote caste polymorphism: (1) nonlinear fitness, $Q(\mathcal{J})$, (2) nonlinear cost constraints, and (3) stochastic effects. As we have already noted, $Q(\mathcal{J})$ can easily be nonlinear as a reflection of economies of scale. Furthermore, it will almost certainly be altered by cooperative effects in task performance. For example, foraging efficiencies are considerably enhanced by recruitment and coordinated work efforts (Wilson, 1971). Similarly, the labor cost (in cal.) of brood rearing are subject to the same economic effects; energetic resources below a certain level will be useless for producing a viable reproductive brood, due to nutritional deficiencies in both workers and reproductives. Of course, whether or not the optimum caste distribution is polymorphic depends on the exact shape of the nonlinearities. A sigmoidal shape for both $Q(\mathcal{J})$ and η_{ij} are probably generic in social insects, but before we can make detailed calculations the form of these functions must be empirically determined. For the remainder of this section we will consider the special case of linear fitness constraints and focus on the effects of stochastic variations in the parameters.

In an uncertain environment the coefficients a_j, which measure the per capita gain by caste j from performing all tasks c,

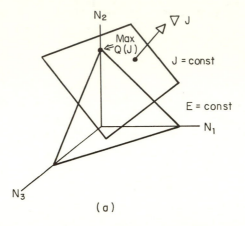

(a)

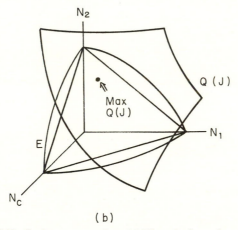

(b)

FIGURE 6.15. Optimum caste mixes. (a) When the fitness function $Q(\mathcal{J})$ and cost function E are linear, the optimum solution is to invest colony resources completely in a single caste. In the example shown here, the maximum $Q(\mathcal{J})$ is obtained if all workers belong to caste 2. (b) When the fitness and cost functions are not linear, the optimum solution will often be a mixture of castes.

should be viewed as random variables. That is, both the task rates T_i and the task efficiencies η_{ij} are subject to random variations about their mean values.

In order to examine the effect of variations in task rates alone, let us assume efficiencies to be constant. Furthermore, in lieu of empirical evidence to the contrary, we shall assume that the T_i are stationary random processes with probability densities $f_i(\tau)$, i.e.:

$$\text{Prob}\{\tau_i < T_i < \tau_i + \delta\tau_i\} \triangleq f_i(\tau)\, d\tau,$$

and that the $f_i(\tau)$ are normally distributed with means $\bar{T}_i$ and variances σ_{ij}^2. Then, because $\mathcal{J}$ is linear in T_i, $\mathcal{J}$ is also normally distributed with mean $\bar{\mathcal{J}} = \bar{\mathbf{a}}^T \mathbf{N}$ and variance $\sigma^2 = \mathbf{N}^T \mathbf{\Lambda} \mathbf{N}$, where $\bar{a}_j$ is the average return from caste j and $\mathbf{\Lambda} = \mathbf{\eta}^T \mathbf{\sigma} \mathbf{\eta}$ is the covariance matrix. The situation is shown graphically in Figure 6.16 for the case of two tasks and two castes. In caste space we have plotted constant risk and return contours. By arguments identical to those in section 6.5, the set of feasible caste ratios is restricted to the segment of the ergonomic constraint boundary labeled as π. The image of π in $(\bar{\mathcal{J}}, \sigma^2)$ coordinates, denoted by $\hat{\pi}$, is shown in Figure 6.16b. It illustrates how risk may be traded off against return.

The tradeoff locus is found as before by solving the quadratic programming problem:

$$\text{Max}\langle \mathcal{J} \rangle + \rho\sigma^2 = \bar{\mathbf{a}}^T \mathbf{N} + \rho\mathbf{N}^T \mathbf{\Lambda} \mathbf{N},$$

subject to

$$\mathbf{c}^T \mathbf{N} \leqslant E, \qquad \mathbf{N} \geqslant 0,$$

for each feasible value of the risk parameter, ρ. The value of ρ is determined by the ergonomic efficiency of alate production, and fixing ρ determines the optimum caste ratio.

The same optimization problem can be stated in a number of equivalent ways, for example: (1) Maximize the mean return $\langle \mathcal{J} \rangle$ while constraining the risk, $\sigma^2 \leqslant$ some preassigned level, k.

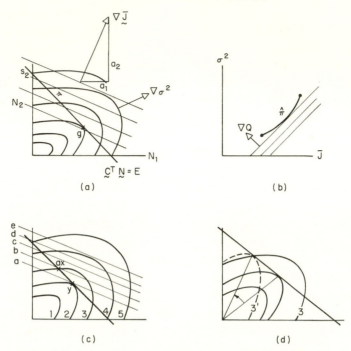

FIGURE 6.16. Risk-return tradeoffs in the evolution of specialized as opposed to generalized castes. See further explanation in the text.

(2) Minimize the risk, σ^2, while maintaining the mean return above a minimum level, k_2. Fixing either the acceptable risk level, k, the minimum profit level, k_2, or the risk-weighting factor, ρ, determines an optimum caste distribution.

Because the constant risk ellipses $\sigma^2 = \mathbf{N^T \sigma N}$ measure the environmental variations, their correlations and the foraging efficiencies, it is not possible to say unequivocally that environmental variability promotes caste diversity. If these caste efficiencies are about the same and the environmental covariances appreciable, this will indeed be the case. For example, consider the situation shown in Figure 6.16c. Fixing ρ determines the optimal caste ratio, located, say, at x corresponding to

222

a risk level of 3 and a return level of b. Now, suppose that the environmental variability increases uniformly so that the risk contour numbered 3 is elevated to 4, 2 becomes 3, and so forth. Then in order to reduce the risk level back to its adapted value of 3 the return must be reduced to a. Moreover, with the same ergonomic constraint the caste proportions must be altered to y, which in this case corresponds to greater caste diversity.

However, an alternate adaptive response can restore the risk level without sacrificing return. This can be achieved in the case given by altering the caste efficiencies η, so as to rotate the risk ellipses, and by restoring the risk level 3' to its former value of 3 (see Figure 6.16d). The inference we draw is that behavioral flexibility of caste members and caste polymorphism are alternate modes of "tracking" a changing environment. As documented earlier, most ant species and virtually all social bees and wasps—but extremely few termite species—appear to have "chosen" increased behavioral flexibility. This may reflect the fact that polymorphism at the level of neural coding is an easier modification to achieve than differential allometric growth during development.

An alternative explanation holds that behavioral flexibility is simply a faster mode of response. In other words, a colony need not await the maturation of new worker cohorts to cope with changing conditions. In this regard it would be well to distinguish between environmental "variability," measured by σ_{ij}^2, and environmental "unpredictability." The distinction is one of time scales. Variations that occur on a time scale shorter than maturation times cannot be countered by developmental responses such as caste specialization unless the colony maintains a large standby reserve in each caste.

Species that employ this strategy would require a more bountiful environment to support what can be characterized as ergonomic inefficiency. In more spartan environments a behaviorally flexible but monomorphic caste system can do as well. Conversely, in an environment where significant variations occur over a time scale of many generations, a caste polymorphism can homeostat the colony without the necessity for a large standby force. Therefore, environmental "variability" is measured by the amplitude variance in the environmental fluctuations, while "unpredictability" refers to the time autocorrelation of the random processes characterizing the environment. We cannot treat time correlations in the context of our static model, but we will have more to say about the distinction in section 6.7.

The foregoing analysis suggests that a useful goal of empirical investigations will be to correlate caste abundance and the degree of caste polymorphism with the magnitude and time scale of environmental fluctuations. Sufficient new data might make it possible to order the magnitudes of the opposing selective forces.

An explicit expression for the interaction between efficiency and variability can be written by referring to Figure 6.17. Caste 2 is the "generalist," capable of performing both tasks T_1

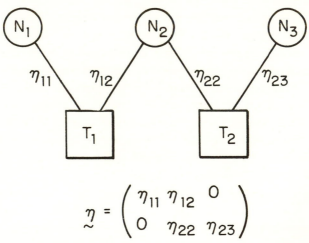

$$\underset{\sim}{\eta} = \begin{pmatrix} \eta_{11} & \eta_{12} & 0 \\ 0 & \eta_{22} & \eta_{23} \end{pmatrix}$$

FIGURE 6.17. The efficiencies of a generalist caste (2) and specialist castes (1 and 3).

224

and T_2. Castes 1 and 3 are specialists. The efficiency matrix thus has the form

$$\begin{bmatrix} \eta_{11} & \eta_{12} & 0 \\ 0 & \eta_{22} & \eta_{23} \end{bmatrix}.$$

Multiplying out the return $\bar{\mathcal{J}} = \bar{\mathbf{T}}^T \mathbf{\eta} \mathbf{N}$ and risk $\sigma^2 = \mathbf{N}^T \mathbf{\eta}^T \mathbf{\sigma} \mathbf{\eta} \mathbf{N}$ expressions, while assuming that the specialist efficiencies are larger than the generalist (i.e., $\eta_{11} > \eta_{12}, \eta_{23} > \eta_{22}$), one can see that the generalists are preferred when the environmental variations are highly correlated.

Although no distinction has yet been made between the various types of tasks, such as foraging, nest construction, and brood care, it is certain that not all tasks are equally vital to the welfare of the colony. Lapses or variations in nest-building efficiency will probably not be as severely penalized as faulty nest repair. Stated another way, some tasks must be performed more reliably than others. In the next section we shall discuss defensive contingencies that must be met with extreme reliability. They are virtually "win or lose" in their outcome: to fail even once could mean the end of the colony. Other tasks, while not life-and-death choices at the appearance of each contingency, nevertheless must be performed correctly a substantial percentage of the time. For example, replete castes ("honeypots") in such ant genera as *Camponotus*, *Myrmecocystus* and *Prenolepis* must provide nourishment for the colony in hard times. Failure to do this greatly erodes colony fitness.

Within the framework of our model, such tasks do not appear directly in the estimates of fitness but rather as additional constraints. The way to quantify their role is discussed in Appendix 6.1. The main result is that if a particular task i must be performed at a minimum rate of $\mathcal{J}_i$ with a reliability level α_i (i.e., $\text{Prob}[\sum_j a_{ij} N_j \geqslant \mathcal{J}_i] > \alpha_i$ where α_i is the probability that task i is performed at a rate $\geqslant \mathcal{J}_i$), then the caste distribution will be polymorphic to a degree which monotonically increases with the required reliability, α_i. For example, if the crucial task can only be performed by a particular caste, then that caste will always be represented in the population by an amount at least $\mathcal{F}^{-1}_i(\alpha_i)$, where $\mathcal{F}_i(\cdot)$ is the cumulative probability distribution of α_i: $\text{Prob}[\mathcal{F}_i < z] = \mathcal{F}_i[z]$.

Thus, the probabilistic constraint is really equivalent to a deterministic constraint that ensures a minimum degree of caste polymorphism. If the task index denotes size classes, then the ith size class will always be represented, but its abundance will generally increase sigmoidally as the reliability level increases. This relation implies the likely existence of a "phase-transition," a threshold below which the caste is not represented.

6.7. DEFENSE OF THE COLONY: WORKERS VERSUS SOLDIERS

Defense of the colony against predators and territorial rivals is one of the most urgent contingencies facing a colony. Indeed, so strong are these selective agents that almost all species have evolved specialized physical and behavioral mechanisms to cope with them. Some of the techniques are very bizarre. Soldiers of the termite genus *Nasutitermes* fire drops of sticky glandular material through snoutlike extensions of the frontal region of the head (Eisner et al., 1976). Soldiers of the termite *Globitermes sulfureus* and workers of the ant *Camponotus saundersi* are "walking bombs": when struggling with enemies they sometimes contract their abdominal muscles in such a way as to tear the body wall open and spray out large quantities of defensive glandular material (Maschwitz and Maschwitz, 1974). Defensive responses can also be narrowly specific. Minor workers of *Pheidole dentata* recruit soldiers much more readily to intruding workers of fire ants and thief ants (which together comprise the genus *Solenopsis*) than they do to any other known kind of enemy. The detection of a single fire ant worker is enough to trigger trail laying and antennal signaling that leads to the enlistment of ten or more soldiers (Wilson, 1975c).

The primacy of colony defense is further indicated by the widespread existence of a true soldier caste among social insects. Almost all termite species have such a caste, while the major workers of ants are more commonly specialized for defense than for any other tasks, including milling and food storage. Soldiers are especially active against invading ants, although those of the army ants (*Eciton*) and driver ants (*Dorylus*) appear to be more effective against vertebrates. During the intervals between battles, the soldiers are mostly or entirely inactive, seldom participating in ordinary labor and engaging in less active patrolling than the minor workers. Also, they rarely comprise more than 20 percent of the worker population; in the great majority of ant and termite species they make up less than 10 percent.

Since soldiers are ordinarily physically larger than their nestmates, they are also proportionately more expensive to create and to maintain. Two independent circumstances can prevent the evolution of a soldier caste. First, the ordinary worker caste might already be well equipped to repel invasions without excessive mortality. This is the case for most ants, which have relatively hard exoskeletons, stings, and defensive glands, and which habitually forage above ground. Second, there might exist larger conditions favoring the colony, such as life in a relatively predator-free environment or confinement within exceptionally well-fortified nests. Each termite species has a characteristic percentage of soldiers in mature nests; this varies from zero in the known species of *Anoplotermes* to as high as 41.8 percent in *Tenuirostritermes tenuirostris* (Haverty, 1977). A comparable amount of variation exists among ant species. No systematic study has been made so far of the possible relation of the soldier/worker ratio to nest structure and to particular features of the environment. Such an analysis might contribute important new insights to the theory of caste evolution.

In order to begin a more theoretical approach, we have devised a model for soldier/worker caste ratios basically similar to previous models concerned with the optimum mix. Balance equations for energy (E), workers (W), and soldiers (S) are written as follows:

$$\frac{dE}{dt} = \text{foraging} - \text{maintenance} - \text{manufacturing} \qquad \text{(energy)}$$

$$= \eta(W)W - [m_W W + m_S S]$$

$$- E\left[\frac{u}{c_W} + \frac{v}{c_S}\right], \qquad E(0) = E_0$$

$$\frac{dW}{dt} = \text{manufacture} - \text{mortality} \qquad \text{(workers)}$$

$$= \frac{uE}{c_W} - \mu(S)W, \qquad W(0) = W_0$$

$$\frac{dS}{dt} = \text{manufacture} - \text{mortality} \qquad \text{(soldiers)}$$

$$= \frac{vE}{c_S} - v(S)S, \qquad S(0) = S_0$$

$$u + v \leqslant 1, \qquad E,W,S > 0,$$

where:

$$\eta(W) = \text{per capita foraging efficiency}$$
$$\mu(S) = \text{per capita worker mortality}$$
$$v(S) = \text{per capita soldier mortality}$$
$$m_W, m_S = \text{mean per capita metabolic rates}$$

The structure of the model is presented in Figure 6.18. As usual, the species is postulated to evolve so as to maximize colony fitness (that is, to maximize Q, the number of new queens produced) subject to the constraints just written. In general, the fitness will be a function of E, W, S, since in order to rear a reproductive brood successfully, energy, labor, and protection

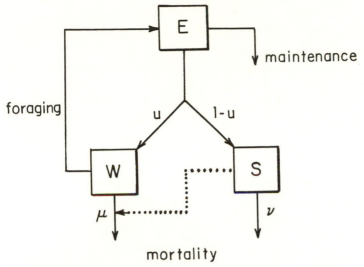

FIGURE 6.18. The energy balance model in the production of workers W and soldiers S.

are all required. In symbols,

$$\overline{Q} = P(S)\mathbf{q}(E,W),$$

where $P(S)$ is the probability that the colony will survive to the end of the reproductive phase. $P(S)$ is the system reliability discussed earlier; it will have the general shape shown in Figure 6.19a, that is, a sigmoid between a lower bound, P_l (the colony may be fortunate enough to survive even without soldiers), and an upper bound P_u. Beyond a certain population of soldiers no improvement in security is possible, but the colony can still be destroyed by a catastrophic event such as flooding or fire, or by an insuperable predator such as an army ant column or pangolin. In this formulation it is assumed for purposes of initial convenience that the manufacturing and metabolism terms are linear.

The soldier mortality, v, will depend not only on S, but on the frequency and severity of the defensive contingencies, which

229

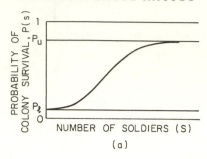

(a)

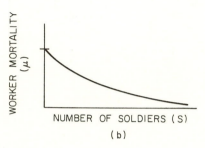

(b)

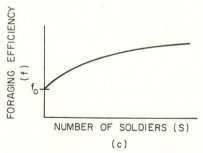

(c)

FIGURE 6.19. The possible effects of the number of soldiers in the colony on (a) the probability of colony survival, (b) the minor worker mortality rate, and (c) the foraging efficiency of the minor workers.

constitute a defense distribution function (DDF). In Appendix 6.3 we present a model for intercolony competition that explicitly accounts for the presence of a nearby competing colony. In the setting of a single colony model v is viewed simply as a random process that inflicts mortality losses on the soldier force.

A reasonable characterization of v is a double stochastic process. Attacks arrive at a random rate, and the amplitudes (measured as the magnitude of the mortality losses) are also randomly distributed. In symbols,

$$v(t) = \# \text{ soldier deaths in } [0,t]$$

$$= \sum_{i=1}^{n(t)} \delta_i(\lambda_2)$$

where $\delta_i(\lambda_2)$ are independent identically distributed random variables with mean amplitude λ_2. Also, $n(t)$ is Poisson distributed:

$$\text{Prob}[\# \text{ attacks in } [0,t] = n] = e^{-\lambda_1 t} \frac{(\lambda_1 t)^n}{n!}.$$

The mean and variance of $v(t,\lambda_1,\lambda_2)$ is $\bar{v} = \lambda_1 t \bar{\delta}_i$ and $\sigma^2 = \lambda_1 t \bar{\delta}_i^2$, respectively. If we normalize the mortality $0 < v < 1$ so that it represents the fractional loss due to attacks, we can write the soldier mortality as $(v_0 + v)S$, where v_0 is the "ordinary" soldier mortality.

The worker mortality, μ, is also a function of the standing soldier force, since workers must fill defensive roles when the soldiers are depleted. The worker mortality is expected to have the general shape shown in Figure 6.19b, which we can approximate in the model by the function $(\mu_0 + \mu e^{-\delta S})$. Here μ_0 is the mortality in the absence of soldiers and δ is the "coefficient of protection." Moreover, soldiers frequently guard foraging lines, so that the foraging efficiency, η, is also an increasing function of S, as shown in Figure 6.19c. It might take the form $f_0(1 - e^{-kS})$.

If we consider the steady state obtained by setting $\dot{W} = \dot{S} = 0$ the optimization problem becomes

$$\mathcal{J}^*(W,S) \triangleq \max_{W,S \geq 0} \{\eta(W,S)W - [m_W W + m_S S]$$
$$- c_W W(\mu_0 + \mu e^{-\delta S}) + c_S S(v_0 + v)\},$$

231

subject to the energy cost constraint

$$G(W,S) = c_W W(\mu_0 + \mu e^{-\delta S}) + c_S S(v_0 + v) \leqslant E.$$

We have plotted contours of $\langle \mathcal{J} \rangle$ (net foraging efficiency) = constant and G (energy loss through mortality) = the (W,S) plane along with constant resource risk contours σ_J^2 (shown in dashed lines) in Figure 6.20. The parameter values were selected arbitrarily in order to illustrate the general nature of the functions.

Consider the mortality cost level labeled G_1 in the figure. The optimum mean rate of energy accumulation $\langle \mathcal{J}^* \rangle$, is

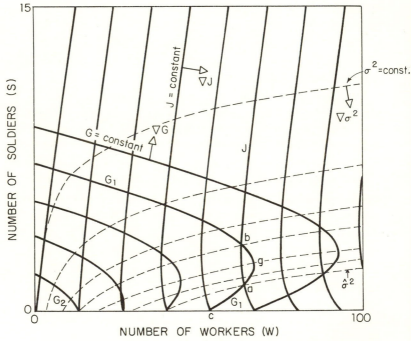

FIGURE 6.20. An illustration of the general relation between net foraging efficiency $\mathcal{J}$; energy loss through mortality G, some of which may be incurred during colony defense; and "risk" or variation through time in attack frequency σ^2. The values provided in this diagram are based on the elementary models previously employed, with arbitrarily chosen parameter values.

located at the caste point g. If the maximum tolerable risk level is $\hat{\sigma}^2$ then the optimum caste point is located on the segment g-b. The segment a-g represents inefficient caste allocations since the same return can be achieved with lower risk along points on g-b. Thus, the point g represents the least number of soldiers for a colony of size G_1, and would be the one selected by a tychophile species. At lower levels of G, corresponding to smaller and/or younger colonies, the optimum caste point may include no soldiers (for example, the contour G_2 in the figure). This is consistent with the earlier conclusion that caste monomorphism can be optimal in the initial stages of colony growth.

The dynamic optimization model is much more difficult to deal with mathematically. Indeed, due to the problems created by nonlinearities and stochastic variation, an analytical solution is beyond reach at the present time. Furthermore, with three-state variables even numerical studies strain the capacity (and budget) of most computer facilities. Therefore, it is probably not worthwhile to attempt a thorough numerical investigation of the model until reliable empirical estimates of the parameters are available. Preliminary numerical investigations using reasonable guesses for the system parameters show that the model has the following properties.

(1) Due to the nonlinearity of the term $e^{-\delta S}$ and the stochastic nature of the parameter v, the optimal strategy is not always "bang-bang" as it was in the reproductive strategy model. Rather, simultaneous production of workers and soldiers is advantageous beyond the initial founding period.

(2) Increasing environmental variability increases the soldier force required to maintain a given risk level. However, if the autocorrelation of the stochastic perturbations is high, suggesting a variable but predictable environment, then a lower soldier force can maintain the same risk level.

(3) The qualitative predictions from the dynamic model coincide with the sequence of static optimization problems shown in Figure 6.20, providing that the fitness does not depend strongly on the *rate* of colony growth and also providing the

environmental autocorrelations affecting $v(t)$ are weak. From Figure 6.20 it can be seen that as the colony grows, the number of soldiers required to maintain a given risk level increases but the proportion of soldiers in the colony actually decreases. This effect will be strengthened if the workers can fulfill defensive roles secondarily.

The optimum caste ratios predicted by Figure 6.20 correspond to maximizing the mean energy accumulation rate $\langle \mathcal{J} \rangle$. For comparison, we have plotted in Figure 6.21 contours of constant mean fitness $P(S)Q(\mathcal{J})$, where $P(\cdot)$ and $Q(\cdot)$ have the generic sigmoid shape discussed above. Several useful conclusions can be drawn. The colony expansion path (labeled e in Figure 6.21) is quite different when $\langle Q \rangle$ rather than $\langle \mathcal{J} \rangle$

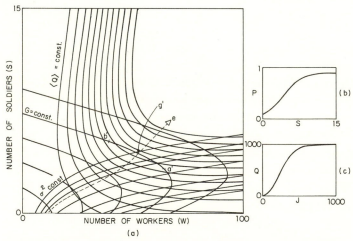

FIGURE 6.21. (a) The general relation between overall virgin queen production $\langle \overline{Q} \rangle$, energy loss through mortality G, and variation through time in energy resources, σ^2. This is the same kind of plot shown in Figures 6.16 and 6.20, except that the risk and return contours are more complicated. The contours depend on the relation given in diagram b between probability of colony survival P and number of soldiers S, and on the relation given in diagram c between virgin queen production Q and net foraging efficiency $\mathcal{J}$. An example of an expansion path of the colony's optimum caste mix is labeled e.

234

is the fitness criterion. Maximizing $\langle Q \rangle$ ensures a higher soldier force, and soldiers begin to be manufactured earlier in the colony cycle. The feasible caste compositions are restricted: the points corresponding to a, g, b in Figure 6.20 are labeled a', g', b' in Figure 6.21. We see that as the colony grows, it is possible for an increasing rather than a decreasing fraction of the colony population to be soldiers.

One final point concerning dynamic versus static optimization models in evolutionary ecology generally is worth keeping in mind. One of the basic properties of adaptive responses is that they tend to neutralize the selective forces that created them. Therefore, if a system is at an evolutionary equilibrium it may be very difficult to retrodict the relative magnitudes of the selective pressures that acted on the system. In a word, evolutionary equilibria forget their histories. For example, if a species has adopted a successful specialized defense in response to a predator, then it may appear that the predator plays only a small role in the ecology of the species, since it can be observed to inflict relatively small losses on the population. In general, the direction of evolutionary changes is quite difficult to deduce from observations at a single point in the history of a species. Thus, comparative and paleontological studies are invaluable for the understanding of contemporary ecology. Unfortunately, the social insects have been almost morphologically invariant for millions of years, and their behavioral adaptations are not preserved in the fossil record. We must continue to search for new and ingenious ways to overcome this handicap.

SUMMARY

Elementary programming models predict that the number of castes should evolve to equal the number of tasks, but in nature it is far less. Fewer than one-fifth of the living ant genera contain species with more than a single physical worker sub-caste, and the maximum number of discrete or otherwise easily distinguishable subcastes does not exceed three or four in any known species. Even when temporal castes are added, the number of castes falls short of the total number of tasks. For example, colonies of *Pheidole dentata*, the best analyzed ant species, perform 26 tasks but contain a total of only four physical and temporal castes.

The disparity between the numbers of tasks and castes can be explained in part as an outcome of the properties of the allometric space. The allometric basis of most variation in ants means that individual species only evolve so as to occupy a narrow region in the space. However, as each task point in the space is successfully covered by a caste, the caste may be flexible enough in behavior to cover nearby task points as well (albeit with less efficiency; see Figure 6.1). Through a wide range of conceivable distributions of task points on the allometric space, the number of castes required to meet all essential tasks is predicted to increase only as the logarithm of the number of these tasks. At the same time, the number of necessary castes decreases as an inverse power of the behavioral flexibility of the caste members. One result of the latter function is that the optimum condition of a species shifts from monomorphism (one physical caste) to polymorphism (multiple physical castes) with a very small decrease in behavioral flexibility. This potentially rapid transition is consistent with the remarkable fact that at least one-third of ant genera with polymorphic species also contain some monomorphic species.

In theory, a variety of constraints can operate to hold the number of realized tasks below the optimum. They are as follows, in the hypothesized order of their relative importance: ergonomic costs, including the energetic expense of constructing large or complex anatomical forms or of placing each caste in the right place at the right time; dispersion in the size of food particles and other environmental variables, making it costly to match each contingency exactly; behavioral plasticity within

236

castes; the overlap in tasks (as expressed in the allometric space model); individual-level selection, which can counteract kin and colony-level selection; the limitations of allometric variation; fidelity costs during development; and in the case of the Hymenoptera, limitations imposed by holometabolous development.

Among the ants, new physical castes have been evolved to serve a limited variety of special adaptations. In particular, major workers are used for defense, foraging, milling seeds, or storing food. This restriction can be used to predict the ratios of the size classes and hence part of the caste distribution function.

Elementary programming models, addressed to the maximization of net energy return in deterministic environments, predict that only a single caste will evolve so as to exploit single and mixed food sources. However, when the sources vary unpredictably through time, multiple castes are favored so as to achieve a compromise between return and risk (Figures 6.10, 6.11, 6.16). A biologically interesting distinction can be made between tychophobe species, which have attained low risk at the price of lowered net energetic yield, and tychophile species, which have "chosen" high yield in exchange for a high level of risk (Figure 6.12).

The risk-return model can be extended to the broader analysis of generalization versus specialization in ecological adaptation, as well as to the evolution of the optimum caste mix to include colony defense.

APPENDIX 6.1. THE OPTIMUM SIZE-FREQUENCY DISTRIBUTION

The optimization problem posed in section 6.4 is

$$\underset{n(s) \geqslant 0}{\text{Max}} \int_0^{s_m} \left\{ [P(s) - C_m(s)]n(s) - k\left(\frac{dn(s)}{ds}\right)^2 \right\} ds, \quad (6.12)$$

subject to the constraints

$$\int_0^{s_m} [C_m(s)n(s) + kn'(s)^2]\, ds \leqslant M \qquad (6.13)$$

$$n(s) \geqslant 0, \qquad 0 \leqslant s \leqslant s_m. \qquad (6.14)$$

In order to cast this in a form amenable to conventional analysis we define the following auxiliary variables:

$$n'(s) = u, \qquad n(0) = 0 \qquad (6.15)$$

$$y(s) = \int_0^s [C(s)n(s) + ku(s)^2]\, ds \qquad (6.16)$$

where $(\cdot)'$ denotes differentiation with respect to s. Differentiating (6.16) yields

$$y' = C(s)n(s) + ku(s)^2, \qquad y(s_m) = M. \qquad (6.17)$$

Thus, the optimization problem (6.12), (6.13), (6.14) can be written in the equivalent form

$$\operatorname*{Max}_{u} \int_{\bar{s}}^{s_m} \{[P(s) - C(s)]n(s) - ku(s)^2\}\, ds, \qquad (6.18)$$

subject to

$$y'(s) = C(s)n(s) - ku^2(s), \qquad y(s_m) = M \qquad (6.19)$$

$$n'(s) = u(s), \qquad n(0) = 0 \qquad (6.20)$$

$$n(s) \geqslant 0, \qquad 0 \leqslant s \leqslant s_m,$$
$$|u(s)| \leqslant U \qquad (6.21)$$

If we regard $u(s)$ as the control variable we now have an optimal control problem analogous to the reproductive strategy model in Chapter Two. Therefore, if we restrict $u(s)$, the slope of the caste profile, to be bounded, i.e., $|u| \leqslant U$, we can apply the Maximum Principle (Leitmann, 1966) in a straightforward way.

238

The Hamiltonian is

$$H(s) = [P(s)n - ku^2 + \lambda_1 u + (C(s)n + ku^2)\lambda_2]. \qquad (6.22)$$

Thus, the adjoint equations are

$$\lambda_1'(s) = -\frac{\partial H}{\partial n} = -[P(s) - \lambda_1 c(s)] \qquad (6.23)$$

$$\lambda_2'(s) = -\frac{\partial H}{\partial y} = 0. \qquad (6.24)$$

The optimal trajectory must be transverse to the terminal manifold $y(s)_m) - M = 0$, which yields the boundary conditions $\lambda_1(s_m) = 0$ and $\lambda_2(s_m) = \text{constant} \triangleq a$. Thus, the equations (6.23) and (6.24) yield

$$\lambda_2 = a \qquad (6.25)$$

$$\lambda_1 = \int_{\bar{s}}^{s_m} \{[P(s) - \lambda_2 C(s)]\} \, ds. \qquad (6.26)$$

The Hamiltonian will have an internal maximum when

$$0 = \frac{\partial H}{\partial u} = \lambda_1 - 2k(1 - \lambda_2)u \qquad (6.27)$$

or $\qquad u(s) = \dfrac{\lambda_1}{2k(1 - \lambda_2)}$

$$= \frac{1}{2k(1 - a)} \int_0^{s_m} \{[P(s) - aC(s)]\} \, ds \qquad (6.28)$$

Then, since $u = n'$, the optimum caste profile is given by

$$n^*(s) = \frac{1}{2k(1 - a)} \int_0^{s} \int_{\bar{s}}^{s_m} \{[P(t) + aC(t)]\} \, dt \, d\bar{s} \geq 0, \qquad (6.29)$$

where $a < 0$ is determined from the normalization condition

$$y(s_{\max}) = M.$$

239

From the above expression we see that the optimal solution is not defined as the fidelity costs $k \to 0$. The reason for this is clear if we examine a simpler finite dimensional version with no fidelity costs:

$$\mathcal{J}^* = \operatorname*{Max}_{\mathbf{n}} \sum_i [P_i - C_i] n_i,$$

subject to

$$\sum n_i = \mathcal{N},$$

where i is the size class index. This is just a linear programming problem whose solution is at a vertex corresponding to a single size class. A fidelity cost term warps the cost function $\mathcal{J}$ such that a mixed caste structure is possible. Analogously, in the infinite dimensional case, neglecting costs means that the ideal caste profile will be concentrated at the size corresponding to the peak of $R(\hat{s})$.

Using the same technique we can add additional performance constraints on the caste distribution. For example, suppose that certain tasks, T_i, $i = 1, 2, \ldots, K$, must be performed at minimum rates $b_1, b_2, \ldots, b_K$, respectively. That is, we must optimize

$$\mathcal{J} = \int_0^{s_m} \left\{ [P(s) - C(s)] n(s) - k n'(s)^2 \right\} ds,$$

subject to the constraints

$$\int_0^{s_m} \eta_i(s) n(s) \, ds \geq b_i, \qquad i = 1, 2, \ldots, K.$$

If the costs are not taken into account then it is easy to show that the optimum caste distribution is a sequence of K "spikes" located at the peaks of each task distribution. That is, the ideal caste distribution corresponds to one monomorphic caste for each size-specific task. Thus, the theoretical maximum number of castes equals the number of tasks—an intuitively obvious conclusion.

APPENDIX 6.2. OPTIMIZATION WITH UNCERTAIN PARAMETERS

For purposes of reference we will summarize here some elementary facts from probabilistic programming. A detailed treatment can be found in Gottfried and Weisman (1973) and Vajda (1972).

The static deterministic programming problem of maximizing a function (e.g., fitness) $Q(\mathbf{N})$ subject to constraints $\mathbf{G(N)} \leqslant \mathbf{K}$, $\mathbf{N} \geqslant 0$ can be written in the form

$$\text{Max}_{\mathbf{N}} L(\mathbf{N}, \xi) = \text{Max}_{\mathbf{N}} [Q(\mathbf{N}) - \xi^T[\mathbf{G(N)} - \mathbf{K}]]$$

subject to the constraints

$$\frac{\partial L}{\partial \xi} \geqslant 0, \qquad \mathbf{N} \geqslant 0$$

where ξ is a vector of Lagrange multipliers. The optimum point, N^*, can also be found by solving an equivalent minimization problem

$$\min_{\xi} \ L(\mathbf{N}, \xi),$$

subject to

$$\frac{\partial L}{\partial \mathbf{N}} \leqslant 0, \qquad \xi \geqslant 0.$$

The dual variables ξ have the following interpretation:

$$\xi_j^* = \frac{\partial Q^*}{\partial K_j}.$$

That is, at the optimum caste point, ξ_j measures the change in fitness due to fluctuations in the jth constraint. For example, $\dfrac{\partial Q^*}{\partial E}$ is the change in fitness due to a small change in the colony's energy resources.

$$\frac{\partial \log Q^*}{\partial \log E_j} = \frac{E}{Q^*} \xi_j$$

is the percent change in fitness per percent change in energy.

For simplicity consider the case when the fitness $Q(\cdot)$ and the constraints $G(\cdot)$ are linear (so that fitness $\equiv$ task performance rate). Then the dual optimization problems are

(a) Maximize fitness:

$$\underset{\mathbf{N}}{\text{Max }} \mathbf{a}^T \mathbf{N},$$

subject to the constraints $\mathbf{CN} \leqslant \mathbf{K}$ and $\mathbf{N} \geqslant 0$. (In section 6.6, $\mathbf{K}$ is just the single energy constraint, E.)

(b) Minimize the sensitivity of the optimum fitness to perturbations:

$$\underset{\xi}{\min} \, \xi^T \mathbf{K},$$

subject to the performance rate constraints $\mathbf{C\xi} \geqslant a$.

There are several approaches to account for randomness in the parameters; the method we have adopted in our modeling is to replace the problem with an approximate deterministic one by employing only the first two moments of the underlying probability distributions.

When discussing performance rate constraints it is easier to consider the dual optimization problem (b). First consider the case where a performance constraint a_{ij} is a random variable with a distribution function

$$\text{Prob}[\textstyle\sum c_{ij}\xi_i \geqslant a_i] = 1 - F_{ij}(z_i)$$

$$\text{where } z_i \equiv \sum_i c_{ij}\xi_i. \ F(\cdot)$$

is typically sigmoidal, as shown in Figure 6.22a. Thus, we envisage the constraint fluctuating about a mean value, and we want to ensure that the constraint is fulfilled with some degree of reliability, α_i, that is:

$$\text{Prob}[\textstyle\sum c_{ij}\xi_i \geq a_i] \geq \alpha_i.$$

That is,

$$\sum_i c_{ij}\xi_i \geq F^{-1}_i(1 - \alpha_i).$$

Thus, the equivalent probabilistic constraint is still linear, but is displaced outward by an amount which increases sigmoidally with the reliability level, α (see Figure 6.22). If $F(\cdot)$ is normally distributed, then one can avoid computing $F^{-1}(\cdot)$ by employing the mean and variance; and using as the constraint

$$\sum_i c_{ij}\xi_i \geq \bar{a}_{ij} + k_i\sigma_i^2,$$

where $k_i = k_i(\alpha_i)$.

If the constraint coefficients, c_{ij}, are random variables, the situation is more complicated, and the deterministic equivalent constraint is nonlinear. We have assumed throughout our discussion that all random variables are normally distributed; the probabilistic constraint is then quadratic. To see this, consider

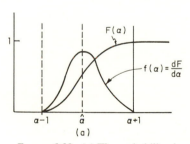 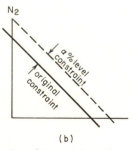

(a) (b)

FIGURE 6.22. (a) The probability density function, $f(\alpha)$, and the probability distribution function, $F(\alpha)$, governing the fluctuations in the performance parameter, a_{ij}. (b) The deterministic constraint equivalent to the probabilistic constraint

$$\text{Prob}[\textstyle\sum c_{ij}\xi_i \geq a_i] \geq \alpha_i.$$

the jth constraint

$$\text{Prob}[\sum c_{ij}\xi_i \leqslant K_i] \geqslant 1 - \alpha_i.$$

$z_i \equiv \sum c_{ij}\xi_i$ is normally distributed with mean $\bar{z}_i$ and variance $\sigma_i^2 = \mathbf{N}^T\mathbf{\sigma}\mathbf{N}$, where $\mathbf{\sigma}$ is the covariance matrix of the c_{ij}. Then the constraint can be written in standard form as

$$\text{Prob}[z_i \leqslant K_i] = F\left[\frac{K_i - \bar{z}_i}{\sigma_i}\right],$$

where $F(z_i)$ is the cumulative distribution function of the standard normal distribution. Let k_i be the constant such that

$$k_i = F^{-1}(1 - k_i).$$

Then $\text{Prob}[z_i \leqslant K_i]$ is equivalent to

$$\frac{z_i - \bar{z}_i}{\sigma_i} \geqslant k_i.$$

Thus, the equivalent deterministic constraint is

$$\sum \bar{c}_{ij}\xi_j + k_i\mathbf{N}^T\mathbf{\sigma}\mathbf{N} \leqslant K_i.$$

The original linear probabilistic program is, under these approximations, equivalent to a quadratic programming problem. In general, this yields only to numerical solution.

APPENDIX 6.3. A MODEL FOR THE EFFECT OF INTERCOLONIAL COMPETITION ON CASTE RATIOS

It is at first tempting to treat intercolony competition from the viewpoint of conventional competition theory by writing Volterra-Lotka-like equations for colonies. This is not likely to be fruitful, however, because many of the crucial adaptations to competition are visible only at the level of the internal structure of the colony, and in particular with reference to the worker/soldier caste ratios. Such models will undoubtedly be complex, and must await a sufficient empirical base to implement numerical solutions. Nevertheless, we felt it worthwhile to at least formulate one such model in the hope that it will stimulate others to pursue the theory.

Consider the case of two colonies competing for a limited resource, R. Each colony can allocate a fraction $0 \leqslant u_i \leqslant 1$ ($i = 1,2$) of its store of energy, E_i ($i = 1,2$) to the manufacture of workers, W_i ($i = 1,2$) and the remaining $1 - u_i$ to the manufacture of soldiers, S_i ($i = 1,2$). The structure of the model is shown in Figure 6.23. Consulting the figure, we can write balance equations for the state variables E_i, W_i and S_i ($i = 1,2$):

$$\frac{dE_1}{dt} = \text{foraging} - \text{maintenance} - \text{manufacturing}$$

$$= f_1(W_1,S_1,S_2) - [m_W W_1 - m_S S_1] - \frac{u_1 E_1}{C_W} + \frac{(1 - u_1)E_1}{C_S}$$

$$\frac{dW_1}{dt} = \text{manufacturing} - \text{mortality}$$

$$= E_1 \left(\frac{u_1}{C_{W1}} \right) - \mu_1(S_1,S_2) W_1$$

$$\frac{dS_1}{dt} = \text{manufacturing} - \text{mortality}$$

$$= \frac{E_1(1 - u_1)}{C_S} - v_1(S_1,S_2).$$

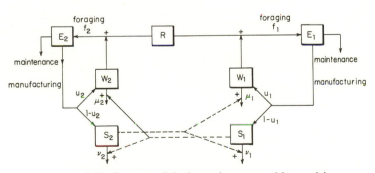

FIGURE 6.23. Structure of the intercolony competition model.

245

Here we have approximated the metabolism and manufacturing rates by linear expressions, but this would probably be too severe an approximation for the foraging and mortality functions. In Figure 6.24 we have sketched the qualitative shape these functions will likely exhibit. Each colony seeks to maximize its fitness, $Q_i(E_i)$, $i = 1,2$, so that the structure of the model is a differential game, and the solution $(u_1(t),u_2(t))$ would prescribe a dynamic Nash Equilibrium (ESS) (see Chapter Three).[4] In the static case, obtained by $\dot{E}_i = \dot{S}_i = \dot{W}_i = 0$, $i = 1,2$, we can plot the fitness contours on the (u_1,u_2) plane and locate the ESS graphically as we did in Chapter Three.

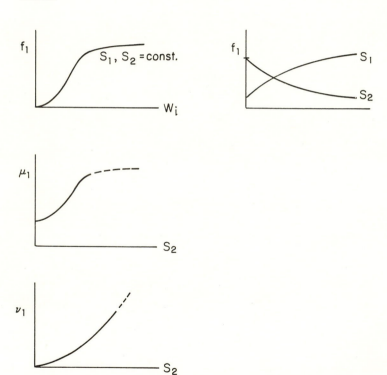

FIGURE 6.24. Possible qualitative shapes of the foraging and mortality functions of two colonies engaged in competition.

NOTES TO CHAPTER SIX

1. If F and S are normally distributed and η_m, η_M are constant, then $\mathcal{J}$ is also normally distributed. It follows that $\langle Q \rangle = Q_0 \int (1 - e^{-\rho J}) e^{-(J - \langle J \rangle)^2/2\sigma^2} \, d\mathcal{J}$. Completing the square in the exponent yields a normal integral whose maximum corresponds to the maximum of the exponent.

2. A risk ellipse $\sigma^2 = \mathbf{x}^T \boldsymbol{\sigma\sigma} \mathbf{x}$ has principal axes of length $1/\sqrt{\lambda_i/\sigma^2}$, where $\lambda_i (i = 1,2)$ are the eigenvalues of the covariance matrix, $\boldsymbol{\sigma}: \lambda_{1,2} = 1/2\{\sigma_F^2 + \sigma_S^2 \pm \sqrt{(\sigma_F^2 + \sigma_S^2)^2 + (2\sigma_{FS})^2}\}$. This major axis makes an angle $\theta = \tan^{-1}[2\sigma_{FS}^2/\sigma_F^2 + \sigma_S^2]$ with the abscissa.

3. We suggest this term from Gr. *tyche*, luck, chance, accident; and Gr. *phobos*, fear.

4. A discussion of dynamic competition games can be found in Mirmirani and Oster (1978).

CHAPTER SEVEN

Caste in the Service of Foraging Strategy

The empirical evidence suggests the existence of four major determinants of caste evolution: (1) the distribution of food items in space and time, (2) the size distribution of the items, (3) the resistance of the items to recovery (including the effectiveness of their antipredator defenses), and (4) the abundance and ecology of competitors and predators. Entomologists do not yet have enough information to conduct an objective analysis of these factors in the form of partial correlations across many species, the method that is becoming increasingly standard in vertebrate ecology. Nevertheless, collecting the required natural history data is likely to be slow and inefficient unless guided by a simultaneously developing theory. With this purpose in mind we propose to force the matter somewhat by making tentative generalizations about the evolution of food habits and its relation to caste. We will stress the features that are unique to social insects and suggest the types of models and field measurements most likely to prove useful in the future. A recent, complementary review by Heinrich (1978) discusses the behavioral and physiological aspects of foraging in the social insects.

7.1. FORAGING TECHNIQUES USED BY SOCIAL INSECTS

Five basic foraging methods can be usefully distinguished:

Type I

Foraging workers leave the colony singly and retrieve prey and other food items as solitary huntresses. Each item is roughly

248

the same size as the forager selecting it—seldom less than one-tenth or greater than several times its body weight. This is the primitive method, practiced by the living nonsocial wasps of the vespoid and sphecoid families most similar to the social bees, wasps, and ants, as well as by the great majority of social wasps. It is also the procedure employed by members of the anatomically primitive ant subfamilies Myrmeciinae and Ponerinae, which are mostly specialized predators on arthropods, as well as by a minority of the species of the phylogenetically more advanced subfamily Myrmicinae. The method can be conveniently referred to as *diffuse foraging*, since the mass of workers appear to diffuse more or less randomly out from the nest.

A closely comparable style typifies the halictine bees and bumblebees. The workers search in solitude, and their body size is correlated with both the size of the flower visited and the weight of the nectar and pollen they are able to carry as individuals.

Type II

Most foraging is conducted in the manner of Type I. However, larger or otherwise more resistant items are added to the resource spectrum through *recruitment*: when solitary foragers encounter items they cannot retrieve alone, they signal the location of the find to nestmates by means of odor trails, ritualized "dances," or some other mode of directional communication. This foraging method is the one practiced by most kinds of ants and higher termites (family Termitidae), by the honeybees, and by some meliponine bees and polybiine wasps. The species employing Type II foraging characteristically select a broader array of dietary items than those utilizing Type I. Ant and termite colonies foraging in the open sometimes recruit soldiers to the food finds, which in turn protect their nestmates and the food itself.

Type III

A notable modification of Type II foraging is the addition of trunk trails. These orientation devices, employed by a minority of ant and termite species, consist of persistent pheromones, which often originate in a different glandular source from that yielding the recruitment pheromones (Hölldobler, 1977). The foragers move away from the nest along the trunk trails, departing at intervals to search over unmarked terrain as single individuals. Some kinds of ants direct their trails to persistent food sources such as populations of honeydew-producing aphids and rich seed falls.

Ant species employing trunk trails are typified by large colonies, well-developed physical caste systems, and some of the broadest diets to be found in the social insects. Examples are numerous among the anatomically more evolved groups, including *Rhoptromyrmex*, *Atta*, *Crematogaster*, *Pheidologeton*, and *Pogonomyrmex* among the Myrmicinae; *Formica* and *Lasius* among the Formicinae; and some of the Dolichoderinae. The habit is also well developed in the harvesting termites of the family Hodotermitidae as well as members of the Termitidae that forage above ground, including the genera *Drepanotermes* and *Hospitalitermes*.

Type IV

A second modification of Type II foraging distinct enough to deserve a category of its own is specialization on exceptionally difficult prey. Examples include the techniques of the termite-raiding ponerine ants; the cerapachyine ants, which so far as we know feed exclusively on the colonies of other kinds of ants; and species of the ant genus *Leptogenys* that hunt sowbugs exclusively. Solitary huntresses locate these objects during forays from the nests and recruit sister workers by means of odor trails; the groups then assault the prey *en masse*.

250

Type V

Army ants engage in "group hunting." That is, instead of searching for food singly the workers proceed out in bands or even entire armies consisting of up to millions of individuals. The group is guided forward by odor trails laid in short segments by members of the van. The leaders soon turn back on their own trails and yield their position to others coming up from the rear. When workers of *Eciton* discover prey, they attract workers differentially in that direction by the use of additional recruitment trails (Chadab and Rettenmeyer, 1975). Group foraging is closely associated with frequent changes in the nest site. Army ants are in fact a functional category defined by the joint display of these two behaviors (Schneirla, 1971). As first demonstrated by W. L. Brown, the association has originated independently in at least three phylogenetic lines of formicids (see discussion in Gotwald, 1971).

7.2. CASTE AND THE EVOLUTION OF FORAGING BEHAVIOR

We will now reconstruct the key steps in the evolution of foraging methods, with special reference to division of labor within the worker caste. The pattern we perceive is dendritic: at each evolutionary grade of foraging, species have had several options in the development of their caste systems. One of the most frequent choices is the avoidance of polymorphism altogether.

The Primitive Foragers

On the basis of anatomical evidence, the living family of nonsocial wasps closest to the ancestry of the ants is either the Mutillidae or the Tiphiidae (Wilson, 1971; Hermann, 1975; Brothers, 1975). In both groups the females are characteristically specialized predators: those of one set of species utilize only tiger beetle larvae, those of another only bee larvae, and so on. This trait is shared by many of the anatomically most primitive living ants. Some species of *Amblyopone*, subterranean forms close to the stem of the Ponerinae, feed exclusively on centipedes. Other specialists among the Ponerinae include the members of *Myopias*, which prey variously on beetles, millipedes, and other ants, and *Discothyrea* and *Proceratium*, which subsist on spider eggs. Most other ponerines accept a wider range of arthropod prey, but as a whole they are not nearly so catholic as many of the phylogenetically advanced ants. Like the extreme specialists, they forage singly and retrieve objects not greatly different in size from their own bodies (Lévieux, 1972). Recruitment is not practiced. This Type I foraging method is shared with some myrmicines, including *Strumigenys* and other Dacetini, and possibly a few specialized predators among the Formicinae such as *Myrmoteras*. The trait might well represent a holdover from the ancestral nonsocial wasps, at least among the Ponerinae.

Type I foraging and dependence on a relatively narrow range of arthropod prey are associated with monomorphism. We interpret prey specialization, whether moderate or extreme, as being one means by which populations avoid competition with other species. Tiny ants such as *Hypoponera* and *Strumigenys* capture the smallest of arthropods, some of which are pursued into the deep penetralia of the soil. These same prey are unavailable to the largest Type I ants, such as *Myrmecia*, *Dinoponera*, and *Odontomachus*, which hunt mostly epigaeic arthropods too large and strong to be captured by smaller ants. We envisage the following chain of evolutionary causation: primitive Type I foraging plus interspecific competition leading to prey specialization, which in turn favors monomorphism.

Specialists on Very Difficult Prey

Some ants considered relatively primitive by anatomical standards have shifted to Type IV or Type V foraging to exploit prey that are exceptionally large or otherwise formidable. Large termitophagous ponerines in the genera *Megaponera*, *Ophthalmopone*, and *Termitopone*, and the *processionalis* group of *Leptogenys* attack termite colonies *en masse*. The raiding parties proceed to their targets along odor trails laid by single scouts; in other words the foraging method is Type IV. The entire ant tribe Cerapachyini consists of species that prey on ant colonies in the same way. The evolutionary early stages of group hunting (Type V foraging) are also associated with specialized predation on ants. This is the mode of life of many of the "true" army ants of the subfamilies Dorylinae and Ecitoninae, including the genera *Aenictus* and *Neivamyrmex*. The amblyoponine genus *Onychomyrmex* subdues large beetle larvae by a form of foraging that is either advanced Type IV or Type V. Other members of the Ponerinae, including some species of *Leptogenys*, employ a rapid form of Type IV recruitment to attack isopods and large beetle larvae.

Specialists on difficult prey are also virtually all monomorphic. We interpret this association to be due, first, to the narrow size range in the prey being exploited by each species, and, second, to the fact that the ants adjust to their victims by behavioral responses based on rapid recruitment rather than by a close matching in size. Even when the prey items are exceptionally large, resistance has been overcome by an increase in the number of attackers rather than by gigantism.

The Advanced Army Ants

The American army ants of the genus *Eciton* and the African driver ants of the genus *Dorylus* represent the extreme development of legionary behavior. The colonies are huge in the case of *Dorylus wilverthi*, some containing over 20 million workers. Masses of foragers sweep over the ground and low vegetation like great vacuum cleaners, seizing nearly all arthropods above the smallest size and even a few more sluggish reptiles and other vertebrates. The worker castes of the various species range from moderately to strongly polymorphic. In *Eciton* the majors function as soldiers and the smallest workers tend to remain within the bivouacs, where they presumably function as brood nurses (da Silva, 1972). But within a large sector of the media class the size of the ants is correlated with the size of the objects they carry (Topoff, 1971).

Thus, it appears that the higher doryline and ecitonine army ants have extended Type V foraging by including most of the size range of arthropods among their prey and even a few small vertebrates. Continuous polymorphism of the extreme monophasic or diphasic types has been added, functioning in part to match the size of the huntress workers to that of their prey.

The Polymorphic Type II *and Type* III *Foragers*

Existing information indicates that polymorphism serves one or the other of two radically different functions in Type II and Type III foragers. Either the workers in the polymorphism series match the objects they carry in size, thus extending the range of objects that can be handled by single workers, or else the largest individuals comprise a soldier caste that defends the minor workers and the food source during the mass retrieval of larger food items.

First we will consider the phenomenon of size matching. In *Solenopsis invicta* and *Veromessor pergandei* the polymorphism is continuous (medias connect the minors and majors), and the workers retrieve objects proportionate to their size (Went et al., 1972; Wilson, 1978). In at least the case of *S. invicta*, workers also carry soil and vegetable particles and handle brood stages correlated with their size. An analogous phenomenon has been noted by Cherrett (1972) in *Atta cephalotes*, a Type IV (trunk-trail) forager. In this leafcutter species the larger medias attack the leaves and flowers containing relatively tougher tissue.

According to Susan A. Gordon (personal communication), the larger workers of *Veromessor pergandei* serve mainly to cut "difficult" seeds, such as those of *Cryptantha*, from their stalks, dropping them to the ground to be picked up and transported homeward by the smaller workers. Within the smaller size class (up to 6 mm body length) there appears to be a correlation between body size and size of the object carried. Continuous polymorphism occurs in some other Type II and Type III ants, including species of *Azteca, Camponotus, Liometopum, Pheidologeton,* and *Proformica,* some members of the *Formica neorufibarbis* and *rufa* groups, and *Pogonomyrmex badius.* Whether it is linked to particle-size discrimination remains to be seen. There is also a need to determine whether such discrimination is the sole or even paramount division of labor in these ants.

One effect of resource matching is to provide the species with a more efficient mode of retrieval over a wider range of food items. The members of polymorphic worker populations are more likely to succeed on the first encounter, singly, and without relying on the recruitment of nestmates, than are the members of monomorphic colonies. If the prey are not diversified—that is, if the ants specialize on arthropods of a limited size array—then the single retrieval rate is also high. Such is in fact the basis of the economy of Type I ants. But where prey of a wide range of sizes and defensive capability are utilized, a higher initial capture rate can be achieved by a continuously polymorphic worker series.

It is an interesting fact that the size-frequency distributions of insects collected randomly in the field are skewed strongly to the large-size classes, assuming a shape that resembles the size-frequency distributions of the workers of continuously polymorphic Type II ant colonies. A typical example from the fire ants is illustrated in Figures 7.1 through 7.3. It can be seen that although insects in the larger size classes are relatively scarce,

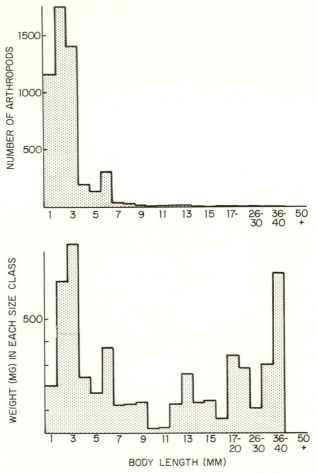

FIGURE 7.1. The upper histogram is the size-frequency distribution of 5,109 arthropods collected at random in dry tropical forest in Costa Rica, while the lower provides the total biomass of arthropods in each size class at the same locality. Although data are insufficient to make precise comparisons with many other habitats, the general forms of the curves shown here are believed to be of wide occurrence. (Based on Janzen and Schoener, 1968.)

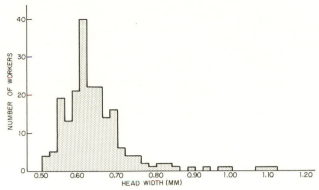

FIGURE 7.2. The size-frequency distribution of 200 randomly collected foraging workers of a laboratory colony of fire ants (*Solenopsis invicta*). A head width of 1 mm corresponds to a body length of approximately 5 mm. The size of the workers is correlated with the size of the items they carry.

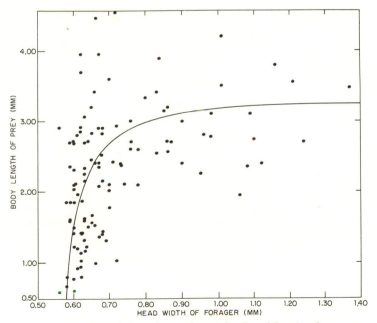

FIGURE 7.3. The relation of prey size to the size of foraging fire ant workers (*Solenopsis invicta*) collecting prey items individually. In this case colonies of fire ants near Tallahassee, Florida, were allowed to attack colonies of termites (*Reticulitermes*) whose nests were broken open and placed near them. (From Wilson, 1978.)

their individual biomass is partly compensatory, so that these classes actually provide substantial amounts of energy for successful predators.

It is possible that the continuously polymorphic ant species have evolved their size-frequency distributions to approximate the size-frequency distributions of their prey. A second factor diminishing the representation of the larger workers could be the greater energetic cost of their manufacture and maintenance, as considered in Chapters Five and Six. This ergonomic constraint can be expected to depress the proportions of medias and majors below the proportions of biomass in the corresponding size classes of insects. If the Costa Rican and New England data provided by Janzen and Schoener are typical of insect faunas generally (Janzen and Schoener, 1968; Schoener and Janzen, 1968; Janzen, 1973b; see for example Figure 7.1), such a disparity is indeed the case. There is a need for close comparisons of the foraging castes of polymorphic ant species, the size distributions of the prey they capture, and the size distributions of the entire arthropod faunas in the nest vicinity.

Bernstein (1976) has demonstrated the significance of another, surprising parameter in the case of ants adapted to cold climates. Workers of *Formica neorufibarbis gelida*, a high-altitude species found in the Rocky Mountains, are continuously polymorphic, with the minor workers also being much darker in color. Bernstein found that the minor workers forage predominantly in the earlier hours of the day, with the larger individuals becoming more active near midday. She suggests the reasonable hypothesis that the minors are able to start earlier because their dark color absorbs sunlight more efficiently and their higher surface-to-volume ratio permits the body temperature to be raised more quickly. Both of these relationships are known to occur widely in other groups of animals (W. J. Hamilton, 1973). In other words, polymorphism in this species of *Formica* may be a device that broadens the foraging time rather than the diet.

By adding Type II or Type III recruitment, polymorphic ants can extend their reach still further. When prey are above a certain size, it is no longer energetically profitable to create workers large enough to retrieve them by solitary effort. Instead, groups of smaller workers can be quickly assembled to accomplish this task. Yet recruitment remains a slower process than single retrieval, a disparity that could serve as a countervailing selective force in some circumstances. Parallel arguments can be made with reference to Type II and Type III foragers that collect seeds.

To explain why some species have chosen the option of continuous polymorphism applied to resource matching, we have considered the hypothesis of ecological release. It is possible that species expand their prey selection—and physical polymorphism—when they occupy habitats with fewer competitors. Where Type I species are monomorphic specialists due to the constraints of interspecific competition, Type II and Type IV continuously polymorphic species are considered in contrast to have expanded their diets as a result of the relaxation of interspecific competition. *Solenopsis invicta*, for example, is one of a group of species known for its aggressiveness toward other species, which permits its workers to forage in local areas with fewer competitors. The size of the major caste of the weaver ant *Oecophylla longinoda*, which conducts all of the foraging for the colony, is highly variable. The species also dominates the trees it occupies (Hölldobler and Wilson, 1977b). *Pogonomyrmex badius* is the only representative of a widespread North American genus of large harvesting ants in the eastern United States and one of the few ants within its range that collects seeds. In the western United States and Mexico, in contrast, competition for seeds occurs among numerous species of *Pogonomyrmex*, *Pheidole*, and *Veromessor*, whose nests are often in close proximity. *P. badius* is also the only polymorphic member of its genus. It is possible, as Bert Hölldobler has pointed out (personal communication), that the two peculiarities are linked, that *P. badius* has

259

responded to the reduced interspecific competition by an expansion of its diet and a pattern of continuous polymorphism to implement the expansion. Recently, Davidson (1978) has demonstrated that the size variation of workers within colonies of *Veromessor pergandei* is inversely correlated with the number of other species of seed-eating ants that coexist with it locally. This phenomenon appears to be uniquely explicable by the process of ecological release.

Although they are less versatile at harvesting single items than the continuously polymorphic species, other Type II and Type III species have some ability at handling items over a wide size range. It is true that solitary huntresses can directly retrieve only those objects light enough to be dragged or carried by their individual effort. These objects are also most likely to be the ones to which the species is adapted in the details of its foraging pattern and digestive physiology and to which it is ultimately limited by interspecific competition. But such species keep open the option of exploiting still larger items by the use of recruitment. To some extent—if the interpretation offered here is correct—they have responded to interspecific competition by choosing recruitment over continuous polymorphism. Recruitment is also employed by the continuously polymorphic species but over a lesser range of food items.

The main disadvantage to reliance on recruitment is the time it consumes, during which competitors can appropriate the food source and predators take the opportunity to attack the assembled workers while they are in the open. We envisage the relationship between size of the food item, foraging method, and risk as indicated in Figure 7.4. There is an item size above which foragers must rely on recruitment rather than solitary retrieval. The delay and greater exposure required by the recruitment procedure significantly increase the losses from competitors and predators. For continuously polymorphic species, in which medias forage in addition to minors, the critical item size is

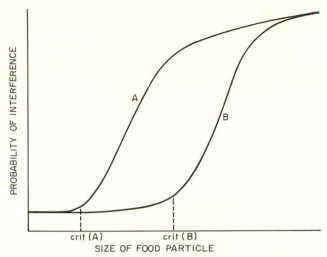

FIGURE 7.4. In ant species with a foraging caste of a single size, represented by curve *A*, colonies must recruit to food items above a relatively small size, *crit A*. During recruitment the probability of interference from competitors and predators is significantly increased and continues to rise as the size of the food item is increased. In species with continuous polymorphism, with medias foraging in addition to minors, a similar curve *B* is obtained but the critical item size is larger.

higher. Species with a monomorphic foraging caste must recruit to a broader range of food objects. To the extent that they utilize larger items they are subject to higher levels of interference during foraging.

Both classes of species can be expected to possess methods for lessening interference during recruitment. Such techniques have been amply documented. *Novomessor albisetosus* and *Paratrechina longicornis* recruit very swiftly and are particularly adept at carrying large objects in groups. Other kinds of ants, such as *Tetramorium simillimum* and the members of *Cardiocondyla*, recruit only in small groups and "insinuate" themselves among larger competitors at food finds. Still another class of species, including the members of *Crematogaster*, *Meranoplus*, *Monomorium*, and

Solenopsis, are slower and more conspicuous, but they are highly successful at repelling enemies with chemical defenses, sometimes of an unusual kind (Hölldobler, 1973).

It is in this context that we can understand the second role of polymorphism in foraging strategies. In place of the behavioral and physiological techniques just cited, some ant species have simply added a soldier caste. When minor workers of *Pheidole dentata* discover a food item too large to carry, they lay an odor trail back to the nest. The trail pheromones attract not only minor workers but also the large-headed soldiers, which cluster around the food source and protect both it and the minor workers from enemies. The role of the soldiers is entirely supportive. In contrast to the medias and majors of some of the continuously polymorphic species, they seldom if ever forage and retrieve food items on their own (Wilson, 1976a).

7.3. CALCULATING THE TASK EFFICIENCY

In order to obtain operational expressions for task performance rates it is necessary to descend below the population level of description to deal explicitly with the behavioral characteristics of individual workers. In this section we shall present a method for calculating task efficiencies from empirical measurements of individual time and energy budgets. Although this is certain to be a laborious procedure, we can see no easy alternative for linking the behavioral and population levels of description. The technique has been applied to one of the simplest examples of task performance: nectar foraging in bumblebees (Oster, 1976). The same procedure can, in principle, be adapted to other species and tasks. The basic requirement is the breakdown of the task in question into discrete behavioral acts, in accordance with the procedure described in Chapter Four.

The first step is to divide the energetic activities of a typical foraging bee into three categories or "states" (a more detailed breakdown is a trivial generalization but would complicate the computations): (1) hive activities between foraging flights, including feeding, thermoregulatory fanning, loafing, etc.; (2) foraging flights, that is, flying to the foraging site and flying between the flowers; and (3) working the flowers—sitting or hovering at the flower and sucking nectar.

Associated with each activity are time and energy expenditure; the overall foraging efficiency depends on how these two resources are budgeted. The forager's activities can be represented schematically by the transition diagram in Figure 7.5. Since we want to account for both energy and time expenditures, we shall associate three probabilistic structures with the transition diagram: (1) the transition probabilities associated with the embedded Markov chain, (2) a "holding-time" structure to keep track of how much time is spent in each state, and (3) a reward structure to keep track of energy profits and losses.

Transition Structure

The probability of moving from state i to state j is $p_{ij}(\tau)$, where the transition probability is conditional on the amount of time, τ, spent in state i. It is assumed that a bee returns to the hive when her honey stomach is filled to its capacity, C_M. Until then she alternates between (1) flying and (2) working the flower.

For now, denote the transition probabilities $P_{12} \equiv P$ and $P_{01} = \hat{P}$; their genesis will become apparent below.

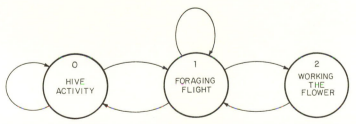

FIGURE 7.5. The three activity states of the foraging bumblebee and their transitions.

Holding-Time Structure

The amount of time spent in each state depends on the joint properties of the bee physiology and the flower distribution and nectar abundance. Denote by $w_1(\tau)$ the probability density function for τ_i, the time spent in state i. For state 2, $w_2(\tau)$ depends on the flower's structure and nectar abundance. Some blossoms require considerable "expertise" on the part of the bee to obtain the nectar reward. Typically, such difficult-to-work flowers have a higher nectar yield relative to the time and energy expenditure necessary to work them (otherwise the bee is likely to desert the species for more generous flower competitors).

The $w_2(\tau)$ is an easily measurable distribution; in the following we shall, for simplicity, employ only

$$\overline{\tau}_2 = \int_0^\infty \tau_2 w_2(\tau) \, d\tau,$$

the mean service time. The complete distribution is required if the effect of varying rewards due to competing foragers is taken into account. For now, we shall assume that all foragers are acting independently (i.e., the number of flowers is very large) and each spends, on the average, τ_2 time units on each flower.

The amount of time spent flying between flowers depends on (1) the bee's velocity, $V(t)$, and (2) the flower distribution, that is, mean interblossom spacing. We shall assume that $V(t)$ is known, or at least $\overline{V}$ (= the average foraging velocity) can be measured.

There are a number of submodels we could use to describe the flower spacing. The simplest assumption is that the flowers are distributed at random. That is, the probability of k flowers in an area A is Poisson distributed:

$$p_k = (\rho A)^k \frac{e^{-\rho A}}{k!}, \tag{7.1}$$

where ρ = mean density of flowers. The flower spacing is then distributed exponentially:

$p(r)$ = probability of two flowers being within
distance r of one another   (7.2)

$= 2\pi r\rho e^{-\pi r^2 \rho}$.

Thus, the mean distance between flowers is $\bar{d} = (1/2)\sqrt{\rho}$ (see Moore, 1954). The "searching radius," R, of the bee is assumed to be a constant characteristic of the species, as is frequently assumed for searching parasites (an example is the Nicholson-Bailey "search area"). The searching time distribution can be computed (Paloheimo, 1971a).[1]:

$$w_1(t) = 2\rho R\underline{V}(t) \exp\left(-2\rho R \int_0^t V(t)\,dt\right). \qquad (7.3)$$

If we use an average foraging velocity, $\bar{V}$, the mean time spent in state 1 is $\bar{\tau}_1 = 1/2\rho R\bar{V}$. Of course, it is unreasonable to assume that the bee alights on every flower that comes within its radius of perception. Therefore, we introduce a probability, P, that a flower detected will actually be alighted on. Thus, the residence time distribution in state 1 must be modified to

$$w_1(t) = 2\rho R\bar{V}Pe^{-2\rho RP\bar{V}t} \qquad (7.4)$$

with mean

$$\bar{\tau}_1 = \frac{1}{2\rho R\bar{V}P}. \qquad (7.5)$$

The probability of alighting on a flower within the radius of perception, P, is the transition probability $P = P_{12}$. (The mean and variance of the more exact residence distribution are computed by Paloheimo [1971b] and can easily be incorporated into the present model.)

If the flower distribution is not random but occurs in patches, a different residence time distribution must be computed. The easiest way to do this is to employ a pair of compounded distributions (Pielou, 1969, p. 87; Johnson and Kotz, 1970, p. 183).

For example, if the flowers are assumed to be distributed in patches such that both the density of patches and the flower density within a patch are Poisson, then a Poisson-Poisson distribution is generated by regarding the within-patch density, $\lambda_2 = A$, as a random variable whose density function reflects the patch distribution:

$$p_k = e^{-\lambda_1} \frac{\lambda_2^k}{k!} \sum_{j=0}^{\infty} \left(\frac{\lambda_1 e^{-\lambda_2}}{j!} \right)^j j^k. \tag{7.6}$$

The mean and variance of the compound distribution are $\lambda_1 \lambda_2$ and $\lambda_1 \lambda_2 (1 + \lambda_2)$, respectively, where λ_1 is the mean patch density. The ratio of the variance to the mean is often used as an "index of clumping"; $I = (1 + \lambda_2)$. Indexes exceeding unity indicate successively higher degrees of clumping, or patchiness (Pielou, 1969, Chapter 8). Note that the pure Poisson distribution has $I = 1$, while the compound Poisson has an index of $I = (1 + \rho A) > 1$. Although the patch density of a flower species is quite likely to be randomly (i.e., Poisson) distributed, the flower distribution within a patch may be much more structured. For example, the blossoms usually cluster on a single plant or a nonrandom distribution around a central focus. Therefore, it may be advisable to employ other compounding distributions, for example, Poisson binomial, or negative binomial Poisson. In any event, this is an empirical decision which must be based on the particular species and situation under consideration. For the purpose of this model we shall adhere to the simplest assumption of a random flower distribution. The results are easily generalized by replacing $w_1(t)$ computed from the appropriate compound distribution.

Remark. The parameter P describing the probability of landing on a flower within the radius (or sector) of perception R has been introduced to account for the fact that a bee does not land on every flower within the search area πR^2 (or the search sector

$\pi R^2[\theta/360]$). If R were known, it would be straightforward to measure P knowing the number of flowers, $\rho\pi R^2$, and the number of landings. Levin and Kerster (1969) have measured the relationship between mean flight distance and mean plant (or inflorescence) spacing and have found the two to be linearly correlated. (A linear relation was also found between the variances.) Thus, the mean flight times, τ_2, and landing probabilities, P, appear to be experimentally accessible quantities.

The residence time in state 0, the hive, also requires some sub-modeling. It is not clear what influences a forager's "decision" to commence foraging or to resume having returned from a trip. Although stochastic effects may intervene, it is not unreasonable to suppose that a forager's proclivity to leave the hive is biased by the success of her previous trip. Returning "empty handed" (with a small net profit, as measured by the honey stomach capacity) would raise her likelihood of remaining in the hive so as not to waste time and energy. Thus, we shall assume that $w_0(t)$ is exponentially distributed with a time constant, λ_0, which is an increasing function of the current net profit rate, g. We shall compute an expression for g once we complete our discussion of the reward structure.

Reward Structure

In addition to the transition and residence time structures $p_{ij}(\tau)$ and $w_{ij}(\tau)$, we must assign a reward rate (positive or negative) to the occupancy of each state and to each transition. Therefore, we define two new distributions on the transition diagram. The first, $r_{ii}(t)$, describes the reward rate while in state i (measured in, say, cal/unit time, or gm nectar equivalent/unit time). The second quantity, r_{ij}, is the energetic cost associated with a transition from state i to state j. The most obvious cost associated with foraging is the power expended in flying from flower to flower. This depends on a number of factors including flight speed, air temperature, and size of the

267

bee. A detailed model of flight energetics is beyond the scope of this book; we shall assume a constant average energy expenditure rate of p_f cal/unit time spent flying for each forager (that is, $r_{11} = p_f$). There is additional cost for each takeoff and landing while foraging (we shall neglect hovering, although it could easily be included). This cost is directly proportional to the size of the bee: for a bee accelerating from rest to a cruising velocity of $\bar{V}$, the "takeoff power" is at least $p_t = (1/2)M\bar{V}^2$, where M is the bee's mass. Thus, a large bee is penalized more heavily for frequent takeoffs and landings than a smaller bee. Indeed, it is observed that larger bees tend to forage on more sparsely distributed flowers which, in turn, offer a proportionately larger nectar reward. We shall assume, for lack of better data, that the takeoff and landing costs are the same:

$$r_{12} = r_{21} \equiv p_t \gtrsim \tfrac{1}{2}M\bar{V}^2.$$

The nectar reward received from each flower depends not only on the flower species but the time of day and the total number of foragers working the flower patch. These considerations will be deferred until later; for now, we simply assume that a constant reward rate r_2 cal/unit time is obtained during each stay in state 2. It is not difficult to allow r_2 to be a slowly varying function of time, $r_2(t)$, to account for the finite blooming time of each flower species. However, since we are working on a time scale of one day, r is left constant. Finally, the energetic costs associated with activities in the hive must be accounted for. Again, an average expenditure rate of r_0 cal/unit time is assumed; a more detailed accounting of hive activities would yield a more accurate assessment. In this connection it is important to bear in mind that the activity level and division of labor change dramatically over a season. Over the period of a day, however, the r_i are approximately constant, and the model can be solved repeatedly for different values of the reward parameters to obtain a seasonal reward profile.

Finally, although the reward from each flower visit was taken as a constant, r, we can allow a stationary phenotypic variation in r by randomizing the final net reward expression, $v(t)$, in the same way as the flower distribution was randomized:

$$v(t) = \int_0^\infty v(t|r)p(r) \, dr, \qquad (7.7)$$

where $p(r)$ is the reward distribution.

In Appendix 7.1 we show how the transition probabilities, $\mathbf{P}$, state residence times, $\mathbf{W}$, and energy budget, $\mathbf{R}$, can be combined to yield an expression for the net caloric profit rate of an individual forager:

$$\eta[\text{cal/unit time}] = \eta[\mathbf{P},\mathbf{W},\mathbf{R}].$$

The expression is complicated, incorporating physiological parameters, P_f, R, M; behavioral parameters, τ_0, P, $\bar{V}$; and resource parameters, r_2, ρ. In Figure 7.6 η is plotted as a function of two of these variables: the mean resource spatial density, ρ, and the size, s, of a forager. This expression corresponds to the phenomenological foraging efficiency $\eta(s,\hat{s})$ employed in the basic models of Chapter Six. In the present case, however, the resource "size" is contained in the reward rate, r_2. A similar model written for an ant of size s foraging on solid food particles would require a specification of two relationships: the fit of the configuration of the worker mandibles to food particles, and the way this anatomical correspondence affects the time and energy required to retrieve food particles to the nest.

7.4. MODELING THE FORAGING PROCESS

In principle, a microanalysis of individual foraging behavior should supply all of the information required to compute the net foraging profit for independently foraging workers. However, a great many species of social insects display some degree of cooperative behavior in their foraging behavior. When

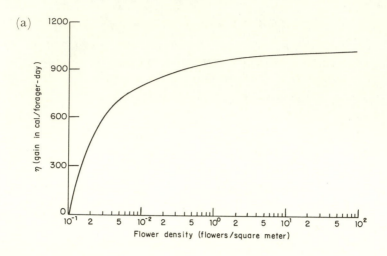

(a)

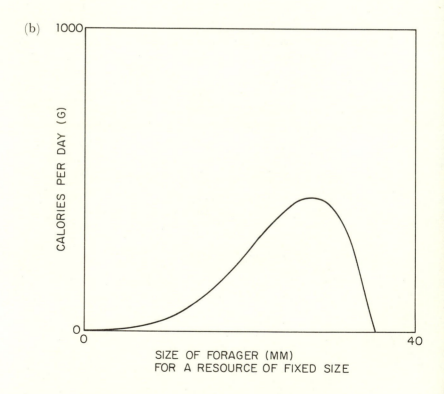

(b)

FIGURE 7.6. (a) The net profit rate of an anthophilous insect forager η as a function of the spatial density of flowers. This curve was derived from the model of foraging efficiency by assuming a Poisson distribution of flowers and by adopting the following values for the constants: average distance (D) to foraging sites = 200 m; mean flight speed v of forager = 1 m per sec; average caloric expenditure during flight, $P_f = 0.009$ calories per sec; capacity of honey stomach, $C_M = 150$ calorie equivalents; average time per flower, $\bar{\tau}_2 = 10$ sec; average time in hive between foraging bouts, $\tau_0 = 600$ sec; maximum foraging bout, $\bar{\tau}_M = 7{,}500$ sec; probability of alighting on a sighted flower, $\hat{p} = 0.5$; caloric expenditure in the hive = $r_0 = 0.0009$ calories per sec; foraging probabilities = $p_{12} = 1$, $p_{21} = 1$; neglect takeoff and landing costs, $1/2Mv^2 = 0$; caloric reward rate from flower, $r_2 = 0.05$ calories per second; foraging day length = 12 hr. (b) The net profit as a function of forager size according to Equation (7.34). In the case shown the foraging efficiency function is skewed implying that larger foragers are more efficient. (From Macevicz and Oster, 1976.)

cooperative effects are important, the semi-Markovian analysis becomes unwieldy, leading to the necessity of a more coarse-grained or phenomenological approach.

One of the simplest but most interesting modes of mass foraging behavior is Type III, based on recruitment back and forth along trunk trails. The following analysis is a modified synthesis of the models by Holt (1955) and Taylor (1977).

By deliberately blurring our vision we can imagine that an ant foraging trail is a fluid stream, which we can describe by the methods of continuum mechanics. Figure 7.7 is a schematic representation of a typical foraging trail, stretching from the nest (at O) to a distant foraging site (at S). Such trails frequently exceed 10 meters in length and end at a fertile resource area such as an aggregation of aphids or extrafloral nectaries. Along the way, foragers leave the main trail, either to follow secondary trails or to forage singly in the adjacent ground cover. In Figure 7.7 we have isolated a small element of the trail and placed a coordinate system on it, labeling the axial distance along the trail as z, while designating as y distances on the axis normal to the trail. Next we denote by $n(x,s,t)$ the number of foragers of size s at position $\mathbf{x} = (y,z)$ at time t. Then the flux of foragers

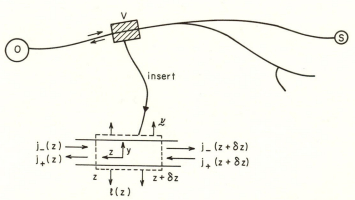

FIFURE 7.7. A foraging trail showing the section over which the balance law is written.

at $\mathbf{x}$ is defined by $\mathbf{J} = n\bar{\mathbf{v}}$, where $\bar{\mathbf{v}}(\mathbf{x},s,t)$ is the average (or root mean-square) velocity of the "fluid" at $(\mathbf{x},t)$. For the time being averages will be taken over all forager sizes, so that we can suppress the dependence on s. A conservation equation on the volume element V may be written as:

$$\frac{\partial}{\partial t} \int_V n \, d\mathbf{x} = \int_{\partial V} \mathbf{J} \cdot \mathbf{v} dA + L(\mathbf{x},t), \qquad (7.8)$$

where ∂V is the boundary of the element V, $\mathbf{v}$ is a unit normal vector pointing out of V, and $L(\cdot)$ is the leakage rate off the trail. If we assume that the foragers who leave the trail do so more or less perpendicular to the trail axis, the above balance law can be written in approximate form as:

$$\frac{\partial n}{\partial t} = [\text{axial flux in } atz] - [\text{axial flux out at } z + \delta z]$$

$$- [\text{radial leakage}]$$

$$= \frac{\partial}{\partial z} n\bar{V} - L(z,t), \qquad (7.9)$$

where $\bar{V}$ is the net axial velocity at position z. If the leakage from the trail per unit length is small compared to the axial flow, we can write the net flux as $\mathbf{J} = j_+ - j_- = n(v_+ - v_-)$, where v_+ is the average velocity of foragers returning to the nest and v_- the velocity of outward-bound foragers.

The formation of the foraging column is complicated in its details. Probably $L \approx 0$ during the initial stages; and Equation 7.8 merely predicts a wave propagating outward in the $(+z)$ direction. For most of the day the trail is expected to be in a steady state $(\partial n/\partial t = 0)$, so that the flux of foragers along the trail is just

$$\frac{d\bar{j}}{dz} = L. \qquad (7.10)$$

Holt (1955) measured the flux and leakage rates of several trail foragers of the red wood ant *Formica rufa* and found that the probability of a forager leaving the trail was nearly constant through time as well as independent of its location along the trail. In other words, ants deserted the trail more or less at random. This corresponds to an average leakage rate per unit length of trail that is approximately proportional to the local density of foragers: $L = -kn$. Equation (7.10) then integrates to

$$J(z) = J(0) \exp\left(-\frac{k}{v}z\right). \qquad (7.11)$$

Holt also found that the average velocity of foragers was nearly independent of z, providing that no obstructions blocked the trail and the temperature remained constant. As a consequence, the forager density along the trail is also exponential:

$$n(z) = n(0) \exp\left(-\frac{k}{v}z\right) \qquad (7.12)$$

His measurements confirmed the validity of Equations (7.11) and (7.12).

When Holt obstructed the *F. rufa* trail at some point along its length the forced detour decreased the convective velocity $\bar{V}$ proportional to flux per unit track width; that is

$$v(\text{obs.}) \doteq v(\text{unobstr.})\left[1 - \text{const.}\frac{J\tau}{w}\right], \qquad (7.13)$$

where τ is the delay and w the track width. He observed no discernible change in trail velocity due to uphill or downhill meanders of the track. However, he did notice an approximately linear increase in trail velocity with temperature and a decrease in trail density, n, as temperature increased. Therefore, the flux must be corrected for ambient temperature T as follows:

$$J(\tau) = n(T)\bar{v}(T).$$

Since radial leakage off the track occurs at a rate proportional to the track density $n(z, y = 0)$, one might view the leakage as a diffusion process. Denote by $n^{(y)}$ the density of foragers leaving a unit length of trail at z, and denote by n_+ the density of foragers returning to the trail laden with a food particle. The leakage process is governed by a diffusion equation:

$$D\frac{d^2 n_-}{dy^2} = F. \qquad (7.14)$$

Here F is the foraging rate during which n_- outgoing foragers are converted to laden foragers: $n_- \to n_+$. A first order model for the function F is produced by treating the foraging process as a mass action reaction:

$$F \sim \hat{k} n_- \rho, \qquad (7.15)$$

where ρ is the density of food particles. The reaction rate constant $\hat{k}$ is expected to be an increasing function of particle size, $\hat{s}$, since larger particles are easier to detect. A more general expression is $\iint d\hat{s}\, ds\, n(s) p(\hat{s}) \hat{k}(s,\hat{s})$, where $\hat{k}(s,\hat{s})$ models the rate at which foragers of size s encounter and pick up particles of size $\hat{s}$. We shall approximate $\hat{k} \sim \hat{k}\hat{s}$, so that Equation (7.14) is rewritten

$$\frac{d^2 n_-}{dy^2} = K^2 n_-, \qquad (7.16)$$

where $k^2 = \tilde{k}\hat{s}\rho/D$. Then the density of outgoing foragers is

$$n_-(y) = kn)(z)e^{-(\tilde{k}s\rho/D)y}. \qquad (7.17)$$

The effective diffusion coefficient, D, is a measure of the spreading rate of foragers away from the trail. A dimensional analysis of Equation (7.16) shows that an average forager in time τ moves approximately $y \sim \sqrt{4D\tau}$ in distance. Consequently, we can regard D as a measure of the "tempo" of foraging activity, an important feature of social organization to be discussed later (section 7.6).

Once a forager picks up a food particle, it presumably makes its way in a more or less directed fashion back to a trail or to the nest. If we assume that laden foragers return to the original

275

trail, we can assign a mean return velocity, $\bar{V}_R$, and write a convective equation for the flux of laden foragers as:

$$\bar{V}_R \frac{dn_+}{dy} = K^2 n_-. \tag{7.18}$$

This states that the returning flux is supplied by the rate at which outgoing foragers assume their burden. Integrating the expression over all y and assuming that the flux vanishes far from the trail, we obtain for the returning flux

$$\mathcal{J}_+(0) = \bar{V}_R n_+(0) = K^2 \int_0^\infty n_-(y)\, dy. \tag{7.19}$$

Using Equation (7.18) we finally obtain

$$\mathcal{J}_+(z) = \frac{k\tilde{k}\hat{s}\rho}{D}\, n(z). \tag{7.20}$$

The total flux of foragers returning to the nest is the sum of those returning from off-trail foraging and those who stayed on the trail until reaching the site s before turning homeward. The former count is obtained by integrating (7.20) over the entire trail length, $z = 0$ to $z = z$:

$$\mathcal{J}_1 = \mathcal{J}_0\, \frac{\tilde{k}\hat{s}\rho}{D}\,(1 - e^{-kZ}) \tag{7.21}$$

At steady state the returning flux from the foraging site s is just equal to the number of foragers that arrive there (assuming no mortality losses):

$$\mathcal{J}_2 = \mathcal{J}_0 e^{-kZ} \tag{7.22}$$

Thus, the net flux of returning foragers is $\mathcal{J} = \mathcal{J}_1 + \mathcal{J}_2$. If the average return from an off-trail forager is C_0 and from a site forager is C_s, the gross return rate to the colony is

$$\mathcal{J}\left[\frac{\text{cal.}}{t}\right] = \mathcal{J}_0 \left\{ C_0\left(\frac{\tilde{k}\hat{s}\rho}{D}\right)(1 - e^{-kZ}) + C_s e^{-kZ} \right\} \tag{7.23}$$

276

To obtain the net profit we must subtract the metabolic costs of foraging. The easiest way to obtain this is to use marked foragers, as did Holt, and measure the distribution of trip times, $f(\tau)$. Since we do not distinguish between forager sizes in this analysis we can multiply the average trip time, $\bar{\tau}$, by an average metabolic rate, $\bar{c}_m$, in order to obtain the foraging cost

$$C = \text{foraging cost} = \left[\bar{c}_m \int_0^\infty tf(t)\, dt \right] \frac{J}{V} \qquad (7.24)$$

If the net profit per trail, $\mathcal{J} - C$, is summed over all trails, an expression is obtained for the net-profit rate of the colony.

Remark. Holt notes that one can obtain an estimate of the total foraging force by the expression:

$$\# \text{ foragers} \sim \bar{\tau} \times \text{flux to or from the nest.}$$

One can easily elaborate this model, for example by treating the size distribution of foragers and resources. In this account, however, we have been limited by the data to a sketch of the general modeling approach.

7.5. RECRUITING AND OPTIMAL FORAGING

The expressions for net foraging profit can easily become quite complicated, so that the optimal strategies must be found numerically. However, by studying a very special case, the general technique can be illustrated. Taylor (1977) carried out a theoretical and experimental study of foraging strategies for the ant species *Solenopsis geminata* and *Pogonomyrmex occidentalis*. In the experiments he considered only trail foraging, and eliminated off-trail leakage by supplying only one or two resource "depots," the characteristics of which he could control. The situation is depicted in Figure 7.8a. Taylor considered two foraging situations: (1) *Solenopsis* foraging on a liquid resource

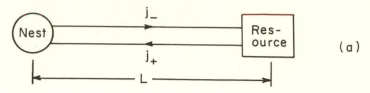

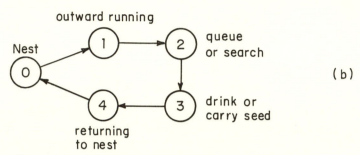

FIGURE 7.8. (a) Nest-to-resource configuration for Taylor's recruiting experiments (based on Taylor, 1978). (b) Transition diagram for the time budget.

and (2) *Pogonomyrmex* foraging on a seed pile. The breakdown of a typical forager's activities is shown in Figure 7.8b; it is a state transition diagram, of the kind discussed in Section 7.3. The activities are: (0) residence in the nest (1) running to the foraging site (2) jockeying for position at the liquid resource (in the case of *Solenopsis*) or searching for seeds (in the case of *Pogonomyrmex*) (3) drinking (*Solenopsis*) or carrying seed at the site (*Pogonomyrmex*), and (4) running back to the nest. To each activity state we can assign distributions of residence times and energy expended (or acquired); transition probabilities between states can also be written. Following the procedures for semi-Markov processes one can then compute the value for the expected net gain rate per forager. However, in order to illustrate the optimal foraging problem Taylor in effect assumed a deterministic process ($p_{12} = p_{23} = p_{34} = p_{45} = p_{51} = 1$) and considered only average values for the time and reward structures. Thus, the time and energy budget is the one shown in Table 7.1. It follows from the definitions in this table that the average net gain rate per forager is

$$g = \sum_{i=1}^{4} c_i t_i \cong c_r \left[\frac{L}{V} + \frac{L}{V'} + t_2 \right] + [\varepsilon - c_3] t_3,$$

while the net profit rate from all foragers is

$$G = j_+ g$$

where j_+ is the returning forager flux ($j_+ = V' n_+$), which in the steady state is just equal to the recruitment flux j_- (neglecting mortality).

In his experiments Taylor presented each species with two foraging sites. By adjusting the distances and rewards of each he could vary the profitability of each source and see how the colony adjusted its foraging strategy to accommodate this inequality. In other words, he could determine whether the ants were adopting an optimal foraging strategy.

If we regard the recruitment flux as the control variable, then the optimal strategy is determined by solving the programming

TABLE 7.1. The time-energy budget of ant foraging (based on Taylor, 1977).

State	Mean Time in State	Mean State Reward Rate
1	$\dfrac{\text{distance to site}}{\text{average running speed}} = \dfrac{L}{V}$	$-c_r[\text{cal/time}]$
2	$\left.\begin{array}{l}\text{waiting time at site } (Sol.) = t_w \\ \text{or searching time at site } (Pog.) = t_s\end{array}\right\} = t_2$	$\left.\begin{array}{l}-c_w \\ c_s\end{array}\right\} \approx c_r$
3	$\left.\begin{array}{l}\text{drinking time at site } (Sol.) = t_d \\ \text{seed carrying time at site } (Pog.) = t_c\end{array}\right\} = t_3$	$\left.\begin{array}{l}\varepsilon - c_d \\ \varepsilon - c_c\end{array}\right\} \varepsilon - c_3$
4	$\text{running back to nest} = \dfrac{L}{V'}$	$c_r' \approx c_r$

279

problem

$$\text{Max} \sum_{k=1}^{P} G_k,$$
$$\small{j^k}$$

where P is the number of available resource patches. If the number of foragers is limited, so that "choice" of forager allocation must be made, then we must add a constraint of the form

$$\sum_{k=1}^{P} \tau_k j^k \leqslant \mathcal{N}, \qquad \mathbf{j} \geqslant 0,$$

where $\overline{\tau}_k$ is the mean total round trip time for patch k

$$\left(\text{i.e., } \overline{\tau}_k = \sum_{i=1}^{4} t_{ik} \right).$$

If there are no "bottlenecks," i.e., all t_{ik} are constant, then the optimal strategy is clearly to put all foragers on the most remunerative source, the one with the largest G_k. That is, since both the total gain and the constraint are linear, a pure strategy is always preferred. Two effects may militate against such a pure strategy: (1) If some uncertainty were associated with each patch, the optimal strategy could easily be mixed. For example, if the reward in state 3 were a random variable, or the transit times, t_i, were stochastic, then maximizing the expected reward subject to reliability constraints would yield a mixed foraging strategy. This case can be handled by the same methods we employed in section 6.4. (2) If a bottleneck occurs at some point in the system, the alteration is equivalent to a "law of diminishing returns" associated with each resource. For example, Taylor assumes that the residence times t_3 are density dependent, i.e., $t_3 = t_3(j_+)$. In the case of *Solenopsis geminata* this is due to congestion at the feeding hole. Taylor assumes a linear dependence of t_3 and j_+, which renders the gains G_i as well as the constraint quadratic, leading in turn to a mixed foraging strategy. In the two-patch case one can solve for the optimal strategy (j_1^*, j_2^*) explicitly (cf., Taylor, 1977).

A more realistic congestion model would treat the bottleneck at the feeding hole as a queue (Kleinrock, 1976). We assume that (a) the ants arrive at the feeding site at a constant Poisson rate $j_- = j_+ = j$; that (b) m ants can drink simultaneously at the source, and (c) the time for an ant to fill up at the drinking hole is exponentially distributed with mean $= t_d$. Then the average waiting time at the feeding site is

$$t_2 = \frac{1}{j} \left[\frac{m^m \rho^{m+1}}{m!(1-\rho)^2} \right] \bigg/ \left[\sum_{n=0}^{m-1} \frac{(m\rho)^n}{n!} + \frac{(m\rho)^m}{m!} \frac{1}{1-\rho} \right]$$

where $\rho = j^{t_d}/m$. (This formula applies whether ants feed in order of arrival, or in random order.) Given measurements of m, t_d and j, t_2 can be found quite easily by graphical means (cf., Shelton, 1960; Ullmann, 1976).

In the case of a seed source, t_2 can be modeled by various functional response equations, such as Holling's "disc model," or by a Nicholson-Bailey random search model.

We have shown that the programming problem to determine the optimal allocation of foragers between two resources is quite complex, even with our simplifying assumptions, so that solutions must be obtained numerically.

In his experiments Taylor was able to demonstrate that both *Pogonomyrmex occidentalis* and *Solenopsis geminata* tend to adjust their foraging patterns in response to resource manipulations in ways that are consistent with the predicted optimal foraging pattern. As usual with strategic models at this relatively crude level of description, there was no indication of the behavioral mechanisms by which the ants actually implement the optimal policy.

7.6. TEMPO AND EFFICIENCY

An interesting fact never before examined by evolutionary biologists is the enormous variation in tempo in activity among different species of social insects. The workers of some ant species walk slowly and with seeming deliberation. As the insects make their rounds among the brood or forage outside the nest, they appear to waste few movements. Examples of such "cool" species include many members of the Ponerinae, within the Myrmicinae most Attini, Basicerotini, and Dacetini, and within the Dolichoderinae, the large members of *Dolichoderus s. str.* In contrast, the colonies of army ants, fire ants, and species of the dolichoderine genus *Iridomyrmex* literally seethe with rapid motion. The workers appear to waste substantial amounts of time canceling one another's actions. One ant in such fast-tempo colonies may run in one direction with a pupa during colony emigration, another in the opposite direction; additional time seems to be wasted by colony members who run back and forth to food sites empty-mandibled.

With the use of the semi-Markov chain description of behavioral sequences it is possible to define tempo quantitatively. Each state (act) has a residence time distribution; from this correspondence one can calculate the distribution of time-to-completion for a particular behavior sequence. That is, from the transition probabilities the probability is computed of traversing a sequence correctly: $r = p_{12}p_{23} \cdots$, providing a measure of individual "reliability" in task performance. Using this component, it is possible to calculate the frequency distribution with which individuals correctly complete a task. The moments of the distribution together constitute a characterization of individual tempo. The performance rate of a group of individuals may increase in a sigmoid fashion with the size of the group as a result of cooperative effects and decreasing returns to scale.

Under certain circumstances tempo can be defined and measured more directly. Thus, in Section 7.4 it was noted that the "diffusion coefficient" D of foragers leaving a trail is a measure of the "tempo" of foraging activity. Hence it is appropriate to use $D \sim y^2/2\tau$ as a measure of tempo, where y is the distance traversed in time τ. In the case of highly active colonies whose members behave more or less independently, one is tempted to make analogies between the roiling masses and the molecular motions of an ideal gas. Thus, the tempo can be defined analogously to the temperature of a gas as the mean "activity level": tempo $\sim \langle MV^2 \rangle$, where M is the average mass and V the average velocity. The mechanical analogy can be pressed still further to define thermodynamic-like aggregation quantities such as entropy. But it is a risky procedure, since the behavior of individual foragers is much more directed and complicated than the random motion of molecules. At best such provisional definitions of tempo inspired by physical models are a first step that in one fashion or another will give way to a microanalysis of the details of insect behavior.

It is our impression that a positive correlation exists between tempo and the mature colony size of species. It was established earlier that a loose correlation exists between the mature colony size and the complexity of the physical caste system. Does a causal relation exist between the three variables? Perhaps species with small colonies must consist of workers that are more "careful," that is, act with greater deliberateness and precision, even at the expense of a lowered rate of productivity. Such species live in circumstances in which colony size is necessarily small for other reasons, such as low density of preferred food or small size of the preferred nest site; under these conditions colony growth would be relatively slow and worker longevity greater. Such colonies are the analogs of K-selected species. The "r-selected" colonies of other species are better fitted to a niche in which the premium is on large mature colony size, rapid colony growth, and a high turnover of workers. Such colonies can afford some degree of inefficiency as the price of their large size and high rate of exploitation of the environment. They are, moreover, "labor saturated" in the sense that each task is attended by many individuals. With such redundancy, the reliability theorems (Chapter One) ensure that the system reliability will be high even if each individual is performing quite erratically.

One coadaptation of r-selection in colonies might be a more differentiated caste system: specialized castes replace the "careful" generalized workers of the low-tempo colonies. Also, since workers of such high-tempo societies live for shorter periods of time, and often will not even survive to the season of the year in which they can produce sons, they will be more likely to lack ovaries. Their specialization into castes will further increase the likelihood of being sterile. These circumstances might account more precisely for the association between large colony size, worker polymorphism, and lack of worker ovaries that has come to light in separate studies.

A graphical version of the same argument is presented in Figure 7.9. Some of the key parameters are indicated in Table 7.1. We recognize that high-tempo ("hot") species are wasteful of energy. Therefore, there must be a premium on high tempo *per se*: high tempo by itself results in a higher level of performance and net energy yield.

Two circumstances are envisioned that lead to the maximization of energy yield at high tempo. First, as suggested in the upper diagram (a) of Figure 7.9, the environment can allow a relatively high gross energy yield as a continuously rising function of colony effort. The most likely circumstance producing this relation is the existence of a relatively rich food source that nevertheless occurs unpredictably in space. The rate of performance in items retrieved per worker in each unit of time will be the product of the frequency of occurrence of the items, the probability of encounter, and the probability of a successful retrieval following each encounter. An increase in tempo will increase the probability of encounter but is likely to decrease the retrieval rate. Such a strategy is most likely to be found in species that capture a wide array of arthropod prey and hence form larger, more polymorphic colonies. In contrast, a species

284

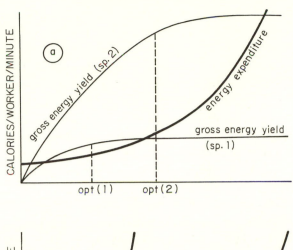

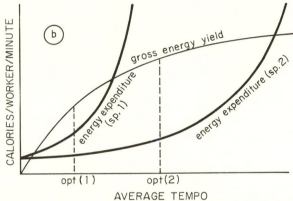

FIGURE 7.9. The evolution of high-tempo ("hot") species is favored by two conditions, depicted in the upper and lower diagram respectively: (a), food is rich but unpredictable in distribution; (b), additional energy losses due to competition and predation are low. These two factors can work singly or in combination.

specialized on a few rare, difficult species of prey is likely to evolve to avoid this strategy and instead adopt a low tempo: the search for the items is careful and aided by special sensory devices, the assault is deliberate, achieved either by stealth or recruitment of nestmates. Foraging, in other words, is more specially tailored to the prey. Similarly, a species adapted to

285

predictable, rich food sources, such as colonies of honeydew-producing aphids, is likely to operate at a low tempo. It needs to invest very little in searching but a great deal in the protection and exploitation of its resource.

The second circumstance leading to a high tempo is a relatively low loss of energy due to the activities of competitors and predators (Figure 7.9b). Some energy expenditure is of course unavoidable due to metabolism, and the expenditure will accelerate as a function of tempo (species 2). If the additional cost due to injury and mortality inflicted by enemies is high, so that the energy expenditure curve rises even more steeply as a function of tempo, the optimum tempo of the colony will be correspondingly lower (species 1). Both curves—gross energy yield and energy expenditure as functions of tempo—can be expected to vary differently between species and hence to produce a wide variation in this single behavioral trait.

SUMMARY

We review certain aspects of the natural history of ants in an attempt to identify some of the selective forces that guided the evolution of caste systems. The key element appears to be food specialization. The kinds of items harvested from the nest environs determine to a large extent the size of the workers, the distances they travel from the nest, the methods by which they retrieve and defend their discoveries, and finally the optimum caste distribution function.

In order to interpret the adaptive significance of caste systems we first set up a classification of foraging techniques used by ants. These range from solitary hunting of specialized prey, an apparently primitive holdover from the ancestral nonsocial wasps, through various modes of recruitment and orientation, to extreme forms such as trunk trails and group foraging. A clearcut relation can be discerned between the foraging techniques of species and the kinds of food they seek.

When caste systems are reexamined with reference to the classification of foraging techniques, their functions also become much clearer. For example, continuous polymorphism serves at least in part to match a wider array of food particles. This adaptation might originate when competition is reduced, allowing ecological release, or when recruitment and assembly results in heavy loss due to attack by enemies.

We analyze the energetics of foraging in a number of simple cases in order to illustrate the steps required to estimate the efficiency of task performance. This procedure, although laborious, forms the link between more traditional studies of physiology and behavioral ecology on the one hand and the evaluation of caste efficiency on the other.

We examine the subject of tempo systematically for the first time and suggest some surprising implications. The workers of some ant species move with extreme slowness and seeming deliberation, while others are swift-moving and apparently less efficient on a per capita basis. This variation appears to be related to differences in diet, and hence colony size and complexity of the caste systems. It can also be related in theory to the mortality risks of foraging workers, especially the risks stemming from attacks by competitors and predators.

APPENDIX 7.1. THE FORAGING EFFICIENCY FUNCTION

The three-state system displayed in Figure 7.5 can be characterized by the following procedure: (1) solve the two-state process shown in the diagram for the net profit rate, g(cal/unit time), while on a foraging trip. (2) The system in the second part of the diagram is then solved using a new mean residence time in state 1 equal to the time to fill the forager's honey stomach to capacity, $C_M(\text{cal})$, that is $\tau'_1 = C_M/g$. In this second computation, formally identical to the first, we allow w_0 and $\hat{p}$ to be (decreasing) functions of τ'_1, the filling time. Thus, we end up with an expression for $v_0(t)$ = net profit at the hive. Then

$$\int_0^{(1\ \text{day})} v_0(t)\ dt \triangleq G$$

is the daily net profit per forager, which will be used in the population equations. The demographic growth rate of the colony will then have been related to the characteristics of the flowers and the foragers.

The transition, reward, and residence time distributions can be given in matrix form as follows:

$$\mathbf{P} = \begin{bmatrix} 1 - P & P \\ 1 & 0 \end{bmatrix} \qquad \mathbf{W} = \begin{bmatrix} \dfrac{1}{2\rho R \bar{V} P} & 0 \\ 0 & \bar{\tau}_2 \end{bmatrix}$$

(a) transition   (b) residence
      structure   time structure   (7.25)

$$\mathbf{R} = \begin{bmatrix} -P_f & -\tfrac{1}{2}M\bar{V}^2 \\ -\tfrac{1}{2}M\bar{V}^2 & r \end{bmatrix}$$

(c) reward structure

There is no difficulty in generalizing the above matrices with each entry as a probability density function, so that the moments as well as the mean value structure of the reward process can be computed.

If a transition rate matrix, $\mathbf{A}$, is defined as $\mathbf{A} = \mathbf{W}(\mathbf{P} - \mathbf{I})$ (Howard, 1971, p. 775), then a set of differential equations can be written for the expected total reward, $v_i(t)$, that the process earns in time t if it is started in state i (Howard, 1971, p. 100):

$$\frac{d}{dt}\mathbf{v}(t) = \mathbf{A}\mathbf{v}(t) + \mathbf{q}, \tag{7.26}$$

288

where

$$q_i = R_{ii} + \sum_{j \neq i} A_{ij}R_{ij} \qquad (7.27)$$

is the "earning rate" of state i. This is a set of linear equations and so the solution can be written down explicitly. However, if we make the reasonable assumption that a bee makes many flower visits per foraging trip, then a simple asymptotic expression can be written for the net earnings, $v_1(t)$:

$$v_1(t) = gt + v_1(0). \qquad (7.28)$$

Here $v_1(0)$ contains the contribution of the initial conditions (i.e., the initial contents of the honey stomach) and a term that accounts for the difference in reward from starting in different states. For simplicity it is assumed that $v_1(0) = 0$; over a day's foraging activity the initial conditions will average to zero, and the process must start in state 1, of course. The quantity g is the "gain" or earning rate of the system in the steady state (cal/unit time). It is computed from

$$g = \frac{1}{\sum_j \pi_j \bar{\tau}_j} \sum_i \pi_i (R_{ii}\bar{\tau}_i + R_{ij}) \qquad (7.29a)$$

(Howard, 1971, p. 868). Here π_j are the limiting state probabilities: computed from $\pi = \pi\mathbf{P}$

$$\begin{aligned} \pi_1 &= 1/\mathbf{1} + p \\ \pi_2 &= p/\mathbf{1} + p. \end{aligned} \qquad (7.29b)$$

Substituting the appropriate values:

$$g = \frac{(-P_f\bar{\tau}_1 - \frac{1}{2}M\bar{V}^2) + P(r_2\bar{\tau}_2 - \frac{1}{2}M\bar{V}^2)}{\bar{\tau}_1 + P\bar{\tau}_2} \qquad (7.30)$$

$$= \frac{(Pr_2\bar{\tau}_2 - [(P_f/2\rho R\bar{V}P) + (1 + P)(\frac{1}{2}M\bar{V}^2)]}{(\frac{1}{2}\rho R\bar{V}P) + P\bar{\tau}_2}. \qquad (7.31)$$

This expression relates the net rate of nectar accumulation

during the actual foraging process to the physiological parameters P_f, R, M; "behavioral" parameters P, $\bar{V}$; and resource parameters r_2, ρ.

Next, in order to compute the net profit at the hive we solve the Markov chain of the second part of Figure 7.10. The only difference from the above calculation is in the values of the matrices in Equation (7.25). The reward associated with state $1'$ is the net reward accumulated in a foraging bout. If we assume that a bee forages until its honey stomach is filled to capacity, C_M, then the residence time in state $1'$ is the filling time $\bar{\tau}_{1'} = C_M/g$ if $\bar{\tau}_{1'} < \bar{\tau}_{M'}$ and $\bar{\tau}_{1'} = 0$ if $\bar{\tau}_{1'} > \bar{\tau}_M$.[2] Thus:

$$\hat{\mathbf{P}} = \begin{bmatrix} 1 - \hat{P} & \hat{P} \\ 1 & 0 \end{bmatrix} \qquad \hat{\mathbf{W}} = \begin{bmatrix} \tau_0 & 0 \\ 0 & \tau_{1'} \end{bmatrix} \qquad .$$

$$\text{(a)} \qquad\qquad\qquad \text{(b)}$$

$$(7.32)$$

$$\hat{\mathbf{R}} = \begin{bmatrix} -r_0 & -P_f \bar{V}/D \\ -P_f V/D & g \end{bmatrix}.$$

$$\text{(c)}$$

Here we have approximated the energy cost of flying to and returning from the foraging site by $-P_f/D$, where D is the

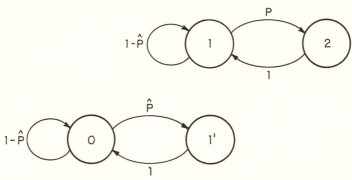

FIGURE 7.10. Decomposition of the 3-state Markov chain (see Figure 7.5) into two 2-state chains that can be solved sequentially.

290

distance of the site from the hive. Then profit rate at the hive is

$$\hat{g} = \frac{[-r_0\bar{\tau}_0 - (P_f\bar{V}/D)] + \hat{P}[g\bar{\tau}_{1'} - (P_f\bar{V}/D)]}{\bar{\tau}_0 + \hat{P}\bar{\tau}_{1'}} \tag{7.33}$$

$$= \frac{\bar{P}C_MH[(C_M/g) - \bar{\tau}_M] - [r_0\bar{\tau}_0 + (1 - \bar{P})(P_f\bar{V}/D)]}{\bar{\tau}_0 + \hat{P}(C_M/g)H[(C_M/g) - \bar{\tau}_M]} \tag{7.34}$$

When Equation (7.31) is substituted into Equation (7.34) the net energetic profit rate per forager is obtained:

$$\hat{g}[\text{cal/unit time}] = f[\mathbf{P},\hat{\mathbf{P}},\mathbf{W},\hat{\mathbf{W}},\mathbf{R},\hat{\mathbf{R}}] \tag{7.35}$$

and

$$\hat{v}_0(t) = gt + v = \text{net profit at hive.} \tag{7.36}$$

Here v is the initial value quantity discussed above. If $\hat{v}_0(t)$ is integrated over a day's foraging activity, the net profit per forager, $G(t) = g \cdot (1 \text{ day})$ can be calculated.

Throughout this section we have assumed that each forager restricts its attention to a single flower type. This is generally true for bumblebees, and so in this respect the model is peculiar to a relatively specialized foraging strategy.

NOTES TO CHAPTER SEVEN

1. This formula applies if $\bar{d} \gg R$. Otherwise, the more complex distribution $w_1(t) = 2\rho R \exp - \rho[\pi R^2 + 2R(\bar{V}t - R)]$ must be employed.

2. The implicit assumption here is that the bee will not waste her time foraging if the gain rate is too low. Otherwise, $\tau_{1'} \to \infty$ as $g \to 0$, a clearly unrealistic situation. Thus $\bar{\tau}_{1'}$ can be written $\bar{\tau}_{1'} = (C_M/g)H[(C_M/g - \tau_M]$, where $H(\cdot)$ is the Heaviside step function at τ_M, and so foraging ceases when $g < C_M/\tau_M$ per forager.

CHAPTER EIGHT

A Critique of Optimization Theory in Evolutionary Biology

Optimization arguments have been the central theoretical tool in our investigations of caste evolution. We now feel obliged to present a critical review of this method, extending the discussion to the broader field of evolutionary biology, in order to evaluate its strengths and weaknesses objectively. Such a review needs to be conducted carefully and repeatedly, because optimization arguments are the foundation upon which a great deal of theoretical biology now rests. Indeed, biologists view natural selection as an optimizing process virtually by definition. This use of the concept of optimality does not, however, require that natural selection create phenotypes "better" than their predecessors in any absolute sense. Rather, we suppose only that physiological design features providing a mortality or fecundity advantage in the local environment will be amplified.

The use of optimization models in biology has its intellectual roots in the development of classical physics during the nineteenth century. The laws of motion in classical mechanics and electromagnetic theory were at first formulated "locally." That is, they were written as differential equations that projected the future trajectory of the system from its present state "one step at a time." Later it was observed that the laws of motion could also be expressed in a global form as extremum principles. This seemed to lend an aura of predesign to the laws of physics, since the solutions to the equations of motion produced trajectories that minimized some quantity. It was easy for some to read divine intent into the evolution of physical laws, such as Fermat's Principle and the Second Law of Thermodynamics, that described nature apparently striving to behave in so economical a fashion.

As the mathematical structure of physical laws was elucidated, the fragility of extremum principles became apparent. Although they lost their metaphysical significance, they retained their esthetic appeal and in many cases still possess the practical advantage of superior computational efficiency.

Joel Cohen has noted ruefully that "physics-envy is the curse of biology." This has been nowhere more true than in evolutionary theory. Biologists have never tired of seeking global designs in nature, but it was not until the advent of modern population genetics that such teleological views were given a concrete form. The crucial step was taken by R. A. Fisher, who showed that the equation governing the change of gene frequencies at a single locus with two alleles under natural selection could be expressed as an extremum principle: the Fundamental Theorem of Natural Selection. The solution to the equations maximized a quantity that Fisher sagaciously called "Fitness." Never mind the delicacy of such a result—constant environment, limited types of density or frequency dependence, no linkage or epistasis, and so forth—teleology had been given mathematical respectability! Efforts to generalize Fisher's result have been ceaseless. They have also been largely unsuccessful, and the validity of the whole enterprise might have been questioned had not a new infusion of optimism come from a completely different quarter.

Analogies between biological evolution and economics had been noted by Marshall and Keynes in the 1920s and by Huxley in the 1930s (Rapport and Turner, 1977). This viewpoint was consistent with the attention then being given competition theory by Lotka, Volterra, Gause, and others, as well as the concurrent development of mathematical population genetics by Fisher, Haldane, and Wright. How much biologists' intuitions about the workings of nature were really influenced by the economic and political currents of the day is difficult to say (but see Lewontin, 1977a). However, to the extent that economic notions penetrated biological thought, so too did optimization arguments, because the entire science of economics deals with the optimal allocation of scarce resources.

The conceptual exchange between economics and biology was reciprocal. Milton Friedman argued in his influential book *Essays in Positive Economics* (1953) that the evolution of business firms proceeds by a process of natural selection for the most efficient. To some economists this conception implies that the present distribution of business economic power is the outcome of inevitable forces. The implications of such a view have caused some politically radical biologists to reject economics-type thinking and hence optimization arguments in order to avoid sullying their scientific investigations with what they consider unsavory assumptions (see, e.g., Allen et al., 1976).

The most recent impetus to optimization modeling in ecology has come from engineering, and especially operations research. Indeed, most of the models we have developed in this book employ mathematical techniques developed to solve problems in engineering and industrial design. The crucial difference between engineering and evolutionary theory is that the former seeks to design a machine or an operation in the most efficient form, while the latter seeks to infer "nature's design" already created by natural selection.

In order to employ engineering optimization models the biologist tries to interpret living forms as in some sense the "best." But just what constitutes superiority is rarely made explicit and is surely seldom very clearly understood. In effect the biologist "plays God": he redesigns the biological system, including as many of the relevant quantities as possible and then checks to see if his own optimal design is close to that observed in nature. If the two correspond, then nature can be regarded as reasonably well understood. If they fail to corre-

294

spond to any degree (a frequent result), the biologist revises the model and tries again. Thus, optimization models are a method for organizing empirical evidence, making educated guesses as to how evolution might have proceeded, and suggesting avenues for further empirical research. At the same time, the concept of optimization can be given a precise definition only in mathematical language. Therefore, it might be a good idea to apply the term only to mathematical models and not to real world phenomena. Otherwise, like the notions of "instinct" and "drive" in ethology, the concept can create more problems than it solves.

In order to justify this pragmatic view of optimization we feel it would be desirable to take a critical look at such models as they are most frequently used in ecology and evolutionary biology. There is little chance that contemporary biologists will repeat the mistake of the early physicists and see in their abstractions the hand of God. But they are still prone to see His surrogate in the automatic processes leading populations to ever higher levels of adaptation. We will show that not even this relatively modest theme can be translated into mathematically sound arguments.

8.1. THE STRUCTURE OF OPTIMIZATION MODELS

The goal of optimization models is to determine the "best way" to allocate a scarce resource among various alternatives. In order to employ quantitative language, that is, mathematics, we must give precise definitions to the innocuous-sounding terms "best way" and "scarce resource." Unfortunately, ambiguity can be eliminated only by introducing a plethora of supporting definitions and concepts. We do not propose to lead the reader through this mathematical labyrinth now; instead, we urge anyone interested in the deeper issues to consult one of the many texts on mathematical programming (e.g., Dorny, 1975; Varaiya, 1973; Intriligator, 1971; Bryson and Ho, 1975;

Leitmann, 1966). The discussion to follow will be brief and largely informal.

The simplest optimization scheme—one employed several times in the present book—is shown in Figure 8.1 and can be

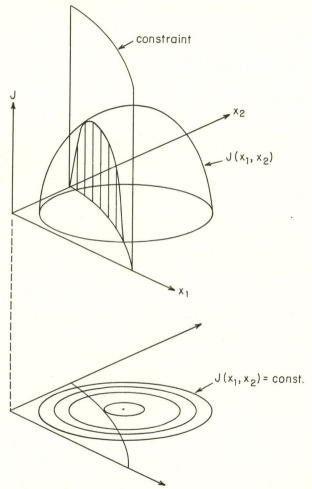

FIGURE 8.1. The simplest optimization scheme.

verbalized as follows:

Maximize $J(x_1,x_2)$, subject to the constraints:

$$f(x_1,x_2) \leqslant \text{constant}; \qquad x_1,x_2 \geqslant 0.$$

This very elementary picture is probably what most biologists have in mind when thinking about optimization models. However, when applied to evolutionary problems it can be seriously misleading, to the point of subverting our intuitions. Let us therefore dissect the picture and discuss some of the complications that intrude during the modeling of evolutionary processes.

Optimization models consist formally of 4 components: (1) a state space; (2) a set of strategies; (3) one or more optimization criteria, or fitness functions; and (4) a set of constraints. Each will now be discussed in turn.

(1) *How do We Define the State of a System?*

A basic goal of virtually all mathematical models is to make as complete a description as possible of the system with reference to the questions being asked. By this we mean precisely the following (Desoer, 1970). Denote by $\mathbf{x}(t) = (x_1,x_2,\ldots,x_n)$ a collection of measurements performed on a system at some time t. If these are sufficient to compute all the future values of the measurements, $x_i(t)$, then the collection $(x_1(t)\ldots,x_n(t))$ are said to constitute a "state description."[1] For example, most population models employ simply the population number $N_i(t)$ $(i = 1,2,\ldots n)$ of each of n species as a state description. It is clear, however, that a census alone is generally not sufficient to predict the future growth of the population. Age structure, nutritional state, sex ratios, and a host of other variables are also crucial for precise population projections.

Among physical scientists there is general agreement, based on generations of empirical evidence, as to what constitutes a state description. For biological populations the situation is not

297

so clear, both because the experimental data are too sketchy and because the complexity of the system is much greater. Therefore, biologists are forced to rely more on intuition and personal experience in selecting the right population descriptors. In this book we have defined the state of a social insect colony by the caste distribution function, $n(t,a,s,x)$, because age, size, and activity distributions seemed to us to be the most distinctive physical features of a social insect colony. Yet the CDF cannot be a complete state description because other phenotypic variables must be important as well, including nutritional state, allometric measurements other than size, and a plethora of behavioral traits and chemical substances about which it is only possible to speculate at the present time. Furthermore, our ergonomic analysis has neglected the genetic structure almost entirely. The models are therefore endemically provisional. They are at best imprecise and will certainly have to be revised with each accretion of empirical evidence. But the situation is merely that which prevails throughout the remainder of evolutionary biology; our state description may be necessary but it is surely not sufficient.

(2) *What are the Strategies?*

The formulation of a mathematical optimization model requires an exhaustive specification of the allowable strategies. That is, the set of strategies must be just as complete as the set of states. For example, in the analysis of reproductive strategies introduced in Chapter Five, the strategic variable was the parameter $0 \leqslant u(t) \leqslant 1$, which determined the fractional allocation of resources to workers as opposed to queens. This requirement makes it very difficult to apply optimization models to evolutionary processes. The reason is clear. The fundamental source of new adaptive strategies is mutation and recombination; natural selection acts only to delete the least "fit" individuals. There is no way to anticipate what new strategies can be generated by these genetic processes: they

298

might include a new enzyme, a modified appendage suited for a new function, a novel behavioral response, and so forth. The very combinatoric richness of the possible molecular permutations in the genetic code makes it impossible to enumerate even a small fraction of the allowable strategies. The essential innovative nature of the evolutionary process precludes an exhaustive list of allowable strategies. The strategy set is always changing, new ones being added and old ones deleted. On the face of it, this property appears to be incompatible with the requirements of optimization models, since the strategy set cannot be specified *a priori*.

Most of the models we have developed to describe caste structure have been economic in conception: they aimed at specifying the optimal allocation of scarce resources among a predetermined set of alternatives. These alternatives consisted mostly of guesses based on our knowledge of natural history. In some sense we had to anticipate the allowable strategic alternatives by reconstructing evolution in our imagination. Our treatment of foraging strategies and caste structure exemplifies the difficulties. By assuming that allometry alone determines foraging effectiveness, we could construct efficiency curves based on a single size parameter and thus establish a relationship between caste polymorphism and ergonomic efficiency (Chapter Six). However, as we pointed out, many species can do equally well with monomorphism simply by employing increased behavioral flexibility and recruitment. Such a strategic alternative would be hard to anticipate with reference to any particular species.

This limitation on mathematical models has been appreciated by workers in the field of prebiotic molecular evolution. The combinatoric possibilities of carbon chemistry make it impossible to enumerate all of the alternative compounds generated by even very low molecular weight substances. Thus, efforts have been initiated to develop theories that can anticipate the most probable reaction pathways from specified initial conditions. At the levels of organismic and colonial evolution such a program is clearly not feasible. However, comparative studies can, to some extent, play the same role. For example, by examining the foraging strategies of a large number of ant species we can get a reasonable idea of the range of strategic possibilities. This has obvious limitations, of course, but in the case of ants at least, the following two conditions are met: (1) there are a large number of comparable species, in fact over 10,000; and (2) they appear to be evolutionarily static, since a large percentage of living genera and even species groups date back to Oligocene times. Thus, we can guess that adaptive radiation has long since exhausted most of the genetically feasible solutions, and we can hope to enumerate a fairly complete strategy set. In other taxonomic groups for which comparative data are scantier, such an assumption is more risky.

It should now be apparent that the central process of evolution—speciation—is relatively impervious to mathematical treatment, since optimization models can only *compare* strategies. The evaluation of qualitatively new adaptive innovations requires most of the answer to be known ahead of time.

Finally, optimization models frequently beg the question of greatest biological interest: how the optimal strategy is implemented. In the reproductive strategy model we ascertained that the "bang-bang" strategy was the best: the colony should produce all of the queens and drones at the very end of each season. Nowhere, however, was the question addressed of how the queen recognizes this optimal switching time. Some sort of biological clock is probably involved, but it was not necessary for our purposes to postulate any particular mechanism. We had only to assume that the genetic potential of the species was sufficient to implement the optimal strategy, whatever the strategy turned out to be. This robust quality is both a strength and a weakness of economic optimization models. On the one hand they circumvent the details of genetic or physiological mechanisms. Yet they force us to assume that at the lower,

300

genetic and physiological levels of organization sufficient flexibility exists to realize all of the strategies.

(3) *What Should be Optimized?*

It is tautological nonsense to say that fitness is maximized: what is fitness? The only unequivocal definition involves maximizing the number of genes projected into future generations. Unfortunately, it is almost always impossible to compute this explicitly in any but the most trivial cases. Hence one is forced to deal with more macroscopic quantities which, presumably, affect genetic fitness in a direct way. In the reproductive strategy model discussed in Chapter Five we used as a fitness criterion the number of alates produced by season's end. There can be little quarrel with this as a fitness measure, all other things being equal. Unfortunately, the life cycle is seldom so clear as in this case, and all other things are seldom equal. The central goal of most of our models has been to see how caste structure could be used as an instrument for increasing the ergonomic efficiency of a colony. Presumably, the greater the efficiency, the greater the reproductive crop and hence the higher the genetic fitness. However, the connection is not so direct as it might seem at first. A colony has other contingencies to meet in addition to energy procurement if it is to maximize its reproductive fitness. One of the most crucial of these contingencies is colony defense; effort expended in that direction must be at the expense of ergonomic efficiency. The situation is typical of optimization models in evolution; there are generally many fitness "components" to be simultaneously maximized if the "overall" genetic fitness is to be maximized. For the purposes of our discussion we can classify the situation according to the number of fitness criteria and the number of "players," or decision makers, as shown in Figure 8.2. Generally, evolutionary optimization models fall into the category of multicriteria optimization or game theory, depending on whether the situation is one of cooperation or conflict between the participants. Team theory

Number of "Players"

		1	2	...
Number of Criteria	1	ordinary optimization	team theory	
	2	vector optimization	game theory	
	:			

FIGURE 8.2. Classification of optimization models by the number of players and the number of criteria.

may have some relevance to models of symbiosis, such as the relation between ants and the aphids they attend.

In our analyses we have learned repeatedly that the conflicting nature of these various criteria prohibits their simultaneous optimization. Risk counters return, specialization on one class of food items counters exploitation of other classes, and so forth. Therefore, maximum genetic fitness can be achieved only by the best compromise solution to these conflicting interests.

When dealing with multiple-criteria decision problems in economics or physics one can frequently use a utility function that establishes a common currency for each criterion, such as dollars or energy. This is seldom possible in evolutionary models, since the "payoff" is in the genes of distant future generations. How can one calibrate present actions by such a standard? We are forced to employ ad hoc methods of combining fitness components into an overall adaptive function. One common method is to combine the separate optimization criteria linearly, or log-linearly by weighting functions that measure the relative importance of each component (Levins, 1968). This procedure is nevertheless usually quite arbitrary, and the assignment of the weighting factors is a difficult problem in itself. In the

302

present book such a procedure enabled us at the very least to classify the life styles of various species according to the relative degree of risk they could accept (Chapters Five and Six). Yet it has to be admitted that the choice of a set of macroscopic fitness functions and of a method for combining them into an overall utility function is another exercise in concealed teleology, and as such it can be no better than our taste and personal experience.

The problem is complicated further by conditions of conflict between species or between individuals of the same species, the game classification in Figure 8.2. There is simply no way of combining the fitness criteria of the various parties, because their genetic interests are basically incommensurable. Therefore, one must examine carefully just what optimization could mean in such circumstances. Optimization is perhaps not even the right word, since nothing is really maximized. We might instead ask how the conflicting interests between the parties can be resolved in a stable fashion. An equitable solution, while desirable in human affairs, is certainly irrelevant from an evolutionary standpoint.

In Chapter Three the worker-queen conflict was diagnosed as a competitive or Nash equilibrium. The idea was to find a set of strategies wherein neither party could gain by changing strategies—provided the other party also maintained the same strategy. However, such a condition is usually only local. That is, the Nash equilibrium is resistant to *small* deviations by either party. When the "players" in the game are populations of individuals, then the Nash equilibrium will usually be evolutionarily stable as well. This is simply because any mutation giving rise to a strategic deviation will always be rare, i.e., be manifest in only a small segment of the population "strategy." When the players are individuals, however, then the Nash point may not be evolutionarily stable. This is because deviations in strategy are produced by variations in phenotypic traits in the players, and these in turn are brought about in part by changes in the underlying genetic properties. For the most part such changes will indeed be gradual and small, and thus the Nash equilibrium may well be stable in the evolutionary sense—an Evolutionary Stable Strategy in the sense of Maynard Smith (1974, 1976). However, we cannot overlook the possibility of a larger-scale strategic change on the part of one player (population) due to a saltational migration or genetic event. Recent work suggests that major phenotypic revisions can occur as a result of regulator gene changes which alter the sequence and timing of developmental events. To the extent that such phenomena play a major role in evolution, global stability criteria will have to be devised to handle large strategic perturbations.

Even if an unequivocal definition of evolutionary stable strategies is devised, there is no guarantee that an equilibrium strategy set can be achieved. That is, achieving and maintaining the Nash point may not be consistent with the laws of Mendelian inheritance. We will return to this question shortly when the role of constraints is examined.

Finally, in addition to the difficulties of the modeling procedure itself, we encounter formidable mathematical obstacles in computing an optimal solution whenever the optimization criterion is made complex to any degree. For example, if the fitness functions are not convex, local maxima can fit together to yield cyclic optimal global strategies (Ekeland, 1977).

(4) *What are the Constraints?*

No biological optimization can be unconstrained. Resources are bounded, and key quantities such as population size and gene frequency must be non-negative. In other words, the optimal feasible solution is always less advantageous than the optimal conceivable solution. Thus, completeness of the constraint set is as crucial as completeness of the state and strategy sets. Neglecting or overlooking a constraint can change the qualitative predictions of the model. Yet to enumerate all of the constraints, much less to write mathematical expressions for them, poses formidable technical problems.

From a mathematical viewpoint, the problem of enumerating all of the constraints is not distinct from that of enumerating the state variables. A constraint means that some of the state variables may be redundant and the problem could have been formulated with fewer state variables. For example, a ball rolling in a bowl only requires two coordinates to specify its position, not three. However, in the model procedures of evolutionary theory it is usually impossible to know *a priori* the theoretical minimum number of state variables required.

For the purposes of the present discussion the constraints can be meaningfully classified as shown in Figure 8.3. The simplest case (the first entry in the classification matrix) has already been adequately captured by Figure 8.1 and needs no further comment. When constraint equations vary in time, the picture becomes much more complicated. For example, reconsider the model discussed in Chapter Two in which the optimal reproductive strategy for an annual social insect colony was estimated. The mathematical structure of the model is summarized in Figure 8.4. In order to maximize the number of reproductives, $Q(T)$, by the end of the season, it is necessary to find the optimal time course of resource allocation, $u(t)$. This allocation must be consistent with a number of static and

	Deterministic	Stochastic
Static	linear or nonlinear programming	stochastic programming
Dynamic	optimal control theory, dynamic programming, calculus of variations	stochastic-dynamic optimization

FIGURE 8.3. A classification of constraints in optimization models. The mathematical techniques used to study them are given in the entries.

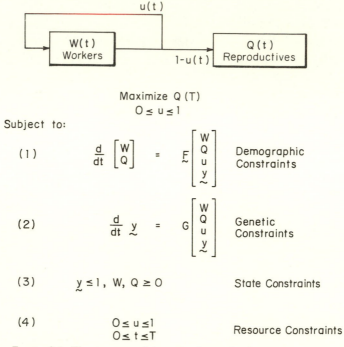

FIGURE 8.4. The structure of the optimal reproductive strategy model presented earlier, in Chapter Two. The vector $\mathbf{y} = (y_1, y_2, \ldots)$ refers to the genetic state variables, and $\mathbf{u}(t) = (u_1, u_2, \ldots)$ are the strategic (control) parameters that replace them in the optimization model.

dynamic constraints. The only dynamic constraints explicitly accounted for in the original treatment were the demographic equations for colony growth. Thus, in realizing the optimal strategy we consciously begged the question of physiological and genetic considerations. In principle, a set of dynamic equations describing the gene frequencies that affect the production of alates should have been appended to the demographic equations. These genetic variables are denoted in Figure 8.4 by the vector $\mathbf{y}(t) = y_1, y_2, \ldots$

Yet a moment's reflection reveals that if we could indeed write the explicit equations governing all the genetic quantities, then there would be no optimization problem to solve. We

306

would then need only to watch the state variables $(\mathbf{x}(t), \mathbf{y}(t))$ evolve in time, and see where natural selection takes the system. The whole purpose of an optimization model is to circumvent the virtually impossible task of writing explicit expressions for "microscopic" details such as gene frequencies. That is, we replace the equations for the microscopic constraints, given by Equations (2) in Figure 8.4, by a "macroscopic" fitness criterion, in this particular case $Q(T)$. However, we must then supply the model with a complete set of control parameters, $u(t)$, that take the place of the genetic equations. This procedure carries with it some strong assumptions. *Most importantly, it requires that the optimal solution be consistent with the genetic constraint equations.* Although this basic assumption underlies virtually all ecological optimization models, there is no guarantee that it is valid.[2]

Indeed, it is not difficult to produce models in which the correspondence assumption is demonstrably false (Slatkin, 1977; Auslander, Guckenheimer, and Oster, 1978). Quite strong mathematical restrictions must be placed on the constraint Equations (1), (2), and (3) in order to characterize their solutions by an extremum principle. The assumption, frequently made, that "sufficient genetic flexibility exists to realize the optimal ecological strategy" is an assertion of pure faith. Rather like the concept of the Holy Trinity, it must be believed to be understood. Let us be clear about this difficulty, since it is central to all optimization models presumed to operate in ecological time. The fitness functions and control parameters must summarize all of the relevant microscopic details not explicitly accounted for in the model. The procedure is analogous to the practice in chemistry of summarizing the aggregate effects of microscopic molecular activity by a macroscopic quantity such as free energy or entropy, which is extremized at the thermodynamic equilibrium. It is easily possible that the microscopic equations are mathematically incompatible with the existence of a macroscopic fitness function. For example, even in the simplest case of 1 locus and 2 alleles, it is known that the genetic equations of motion do not support an extremum formulation except in very restricted situations (Shashahani, 1978). For the most part, ecologically relevant traits will surely be polygenic.

The mathematical treatment of such gene ensembles appears prohibitive at this time. If we regard the system as autonomous and simply follow the evolution of gene frequencies and population, then it is only under very special assumptions concerning the form of the population and genetic dynamics that the system evolves so as to maximize anything. For example, if one assumes the simplest sort of density and frequency dependent selection then one can show that the equilibrium gene frequency configuration corresponds to a (local) maximum of the total population (cf., Roughgarden, 1978). This result, however, cannot be regarded as a general feature of population models (Oster and Rocklin, 1978). Thus it is unlikely that a mathematically convincing link will soon be established between the microscopic dynamics of genotypes and the macroscopic processes of phenotype selection, although attempts in this direction are being made (see, for example, Rocklin and Oster, 1976; Levin and Udovic, 1977).

Another justification for neglecting genetic dynamics frequently invoked during ecological modeling is the supposed disparity of time scales. If one could assert that evolutionary processes involving gene changes occur at rates much slower than demographic changes, then he would be justified in transforming the dynamic equations for gene frequencies into static constraints. In Figure 8.4 this corresponds to setting $\dot{\mathbf{y}} = 0$. In formal terms, it appears that we might then eliminate the genetic variables from consideration by simply solving $\mathbf{0} = \mathbf{g}(\mathbf{x},\mathbf{y},t)$ for $\mathbf{y} = \mathbf{y}(\mathbf{x},t)$, where $\mathbf{x}$ are the demographic state variables W and Q, and substituting for $\mathbf{y}$ in constraint Equation (1). But several things can go wrong with this program. First, the model approximation may not be valid: genetic and demographic processes can be tightly coupled. For example, density-dependent natural selection can alter gene frequencies significantly within a few generations—well within ecological time—and contribute to demographic changes which in turn shift the program of natural selection. This has been demonstrated in theory and actually occurs in populations of insects, rodents, and other prolifically breeding organisms (see Pimentel, 1968; Nolte et al., 1969; Krebs et al., 1973; Auslander et al., 1978). Second, the "static" genetic constraints might not be well behaved. Saltational changes due to gene migration, mutation, or recombination can produce a constraint surface with folds (or "catastrophes" in current mathematical jargon). This phenomenon enormously complicates the demographic dynamics. The likelihood of formulating an optimization scheme that captures such constraints is remote.

Still another pitfall is the possibility that the dynamic equations will exhibit chaotic behavior. Even though the equations may be perfectly deterministic, they can still behave in a wholly unpredictable way, in effect resembling a random process. Such bizarre behavior could prove quite common in ecological models when nonlinear density or frequency dependent effects are important (see May, 1976; May and Oster, 1976; Guckenheimer et al., 1977; Oster and Guckenheimer, 1977). No methods exist at the present time for the solution of optimization models when the constraints exhibit chaotic or discontinuous behavior. It is not even clear how extrema could be mathematically defined in an unambiguous way.

Short of chaotic behavior, the dynamic constraints can exhibit periodic behavior of various kinds. Population cycles are in fact a commonplace in ecology. Such responses generally preclude characterization by an optimization formulation, although cycle averages alone can be meaningful in some cases (see Holmes and Rand, 1977).

Finally, we must confront the problem of stochastic influences in the model equations. No realistic model can afford to neglect them, since all organisms evolve in a more or less uncertain environment. It was shown during our earlier analysis of caste ratios that reasoning with deterministic averages can yield totally erroneous conclusions. When an attempt is made to include stochastic effects the mathematical arsenal is quickly exhausted, because stochastic optimization theory is still at a primitive state. Moreover, there is an important distinction between the definitions of the system "state" in deterministic and stochastic models. The state of a stochastic model is described not by a vector of precise measurements, $\mathbf{x}(t)$, but by a probability distribution. Furthermore, as we discussed in Chapter Five, maximizing ergonomic efficiency is generally *not* the same as maximizing reproductive fitness when stochastic influences are important. This is because the conversion of energy and labor into reproductive alates is expected to be nonlinear. Finally, as we saw in Chapter Three, the fitness maximum may be rather broad and flat. Thus, selection might act—but weakly—in the vicinity of such optima, and one might expect the system to be dominated by rather large random deviations from the equilibrium. Computing the evolution of the system as well as validating the model empirically poses serious mathematical and experimental difficulties (cf., Astrom, 1970; Melsa and Sage, 1973). Some progress has been made in the case of static optimization by employing only mean-variance analysis to construct deterministic equivalent models (Chapter Six). Even there, however, linear models became nonlinear. In the dynamic case the difficulties are enormously greater, and at the present stage we can hope for little more than exhaustive computer studies. These in turn will require large quantities of real data.

The evolutionary success of a species is determined by how successfully it tracks its physical and biological environment. Probabilistic uncertainty enters into the evolutionary process at every level, from randomness generated by allelic segregation, recombination and mutation, through chance processes in mating patterns and ecological selective forces such as weather, climate, predation, and resource abundance. Cohen (1976) has presented some disturbing examples of how these probabilistic processes can generate a fundamentally unpredictable evolutionary process. Such models hint that historical accident plays a more decisive role in evolution than we have been willing to accept. Given our present state of knowledge, a scientist attempting to be scrupulously objective must remain skeptical concerning the potential long-range predictive power of evolutionary and ecological theory.

Yet, for all the pessimism justified by these mathematical arguments, evolution cannot be demoted to the status of a random meander. Order and progression in evolutionary processes are consistently perceived. Nevertheless, we must always bear in mind the crucial fact that evolution is a history-dependent process. Adaptations are not "designed" *de novo* by nature. Rather they are jury-rigged using the material available at the time. Evolution, in the words of Jacob (1977), is a "tinkerer," not an engineer! As systems become more complex, the historical accidents play a more and more central role in determining the evolutionary path they will follow. This is simply because the combinatorial number of possible alternatives multiplies enormously as the complexity of the overall system increases through the addition and increase in size of the subsystems. When we compare the complexity of the eucaryotic genome with that of the hydrogen atom it is little wonder that the physical scientist can perceive the "economy of nature" more clearly than the biologist!

8.2. WHAT IS THE PROPER ROLE FOR OPTIMIZATION MODELS IN EVOLUTIONARY BIOLOGY?

When all of the canonical limitations are taken into account (see Table 8.1), we are left with a more modest view of the role of optimization models in biology. Rather than a grand scheme

TABLE 8.1. The four essential components of optimization models and the difficulties encountered in specifying them in evolutionary biology.

Model Component	Difficulties
1. State space	a. Generally impossible to give a complete objective description.
2. Strategy set	a. More genetic combinations are possible than can be specified.
	b. Usually not possible to specify how the strategy could be implemented.
3. Optimization criteria	a. All tractable definitions are indirect.
	b. Difficult to define a common currency for multiple criteria and a proper global notion of equilibrium in conflict situations.
	c. Severe mathematical difficulties if the optimization criteria are not simple.
4. Constraints	a. Impossible to enumerate all of the constraints.
	b. Constraints may be incompatible with stable optimum.
	c. Dynamic constraints may behave chaotically; static constraints may have "catastrophes."
	d. Mathematical machinery for handling stochastic effects is not well developed.

for predicting the course of natural selection, optimization theory constitutes no more than a tactical tool for making educated guesses about evolutionary trends. If we wish to view evolution as an "optimizing" process, and to retain mathematical modeling as an analytic tool, we are at least forced to admit that an element of teleology has entered the theorizing. This appears to be unavoidable and should be frankly admitted at the outset. The prudent course is to regard optimization models as provisional guides to further empirical research, and not necessarily as the key to deeper laws of nature. The proper role of optimization models, in our view, is to provide the means for recreating short-term evolution in the imagination. To this end a three-step procedure is suggested:

(1) Construct a model on the basis of natural history experience and intuition. A complete model will contain four compo-

nents: a state description, a collection of strategic alternatives, a set of fitness criteria, and a set of constraints.

(2) Work out the logical conclusions of the model.

(3) Compare the conclusions with empirical observations. Thus, the model-building process can be analogous to a blind experiment in the laboratory. Models sufficiently complex to produce nonobvious results are also able to compete with one another, just as one experiment discloses the operation of the crucial process and another does not. If the surviving model then sufficiently matches the available empirical data, the biologist is provisionally justified in feeling that he has captured the essential features of the phenomenon. At the very least his intuition has been tested and enhanced. If the predictions of a particular model are erroneous, he knows precisely how to revise it, since the assumptions were made explicit and clear at the outset. Thus, step by step the investigator proceeds by the "strong inference" method that has proven so effective in the physical sciences and molecular biology (Platt, 1964).

Optimization models, while useful for surmising "strategies," can hardly ever address the more basic question of "tactics." That is, the particular chemical or physiological mechanism a species employs to implement a strategy is contingent on its history. Moreover, an optimization model cannot even find the best of all possible strategies, it can only eliminate many inferior ones from among a preselected set. The only practicable goal is to identify local optima, conditioned by the organism's ecological niche. If it is sound, the resulting theory is at least equally explicit as the inductive formulations of natural history and more vulnerable to falsification. Above all, it is linked to other generalizations by a shorter chain of logical steps. Good theory, being the product of human imagination, is also that which provides the greatest esthetic satisfaction, perhaps because esthetics is to some extent the judgment of formulations that work in the realm of the poorly understood. P. A. M. Dirac has said that physical theories with some mathematical beauty are also the ones most likely to be correct. However, in view of the dominant role of history in determining the awesome complexity of the biological world we cannot as a rule expect mathematical theories in biology to have the same elegant simplicity we find in physical theories.

SUMMARY

In examining a behavioral or physiological trait, it is possible to adopt one or the other of three philosophies: (1) the trait is *a priori* "adaptive," that is, has been molded by particular agents of natural selection; (2) it is a random event; (3) it is partially adaptive and partially an epiphenomenon created by stronger selection occurring on other traits. Much of the present chapter has been devoted to the demonstration that only (3) is supportable. Given the complexity of the eucaryotic genome, the finiteness of population sizes, and the limited number of generations in which evolution can occur, it is simply impossible for selection to independently optimize every aspect of an organism's genotype. But the nihilistic opposite of this conception is equally untenable, unless we are willing to deny the underpinnings of a large part of modern biology.

Some traits are certainly adaptive and have been "optimized" in the literal terms of elementary Neo-Darwinian theory, while others are random manifestations in the sense of being selectively neutral or secondary, canalized consequences of the existing genetic order. In our opinion, the way forward in evolutionary theory is not through the formulation of global statements about the evolutionary process, but through the prudent choice of paradigmatic examples that permit the role of natural selection to be analyzed with unusual clarity. Social insects appear to offer many such opportunities; they are the "squid axon" of evolutionary sociobiology.

In an effort to temper our own strongly selectionist view of the evolution of caste systems, we have in this chapter presented a critique of the use of optimization theory in evolutionary biology. We conclude that the mathematical techniques of the theory cannot be used to make long-range predictions of evolutionary processes. Indeed, the concept of lone optima toward which many species can be said to be moving along certain trajectories appears to be an unsupportable metaphysical notion. Nevertheless, in many cases the course of evolution might be predicted over *short* distances. And if carefully constructed, optimization models can be used to identify local optima and to test adaptation hypotheses in a more rigorous manner than is possible by inductive natural history. The construction and testing of models is a potentially powerful technique analogous to blind experimentation conducted in the laboratory.

Good optimization theory requires a systematic accounting of four components of its models: a state space, incorporating all of the relevant variables; a set of conceivable strategies; one or more optimization criteria, or fitness functions, precisely defined; and a set of constraints that limit the approach to the idealized local optima. Biological systems create special difficulties for the characterization of each of these four components.

In spite of the difficulties, we feel justified in making optimization theory the cornerstone of caste theory. The rigidity of insect caste systems, their stability through evolutionary time, and the existence of literally thousands of species that can be examined as independent evolutionary experiments make the theoretical enterprise feasible. If the effort is successful, a deeper understanding of social evolution in the insects will be gained and general evolutionary theory can be additionally tested and refined.

NOTES TO CHAPTER EIGHT

1. A more general definition of "state" is required for systems where stochastic effects are important.

2. Hidden in most optimization models is an assumption that the underlying genetic equations are close to a "gradient flow," i.e., that the system is asymptotically stable in a global sense (Oster and Rocklin, 1978).

CHAPTER NINE

Unsolved Problems

Because science is supposed to consist of the pursuit of soluble problems, scientist authors naturally emphasize positive results, upheld hypotheses, and proven theories. Casual readers are unintentionally given the impression that the only problems left unsolved are either trivial or excessively hard. We wish to close this book with the opposite impression, by explicitly identifying some of the key theoretical and experimental problems that are the most obvious to us. The following catalog is very incomplete. It can be multiplied several times over from a close examination of the text alone, and it will no doubt grow dendritically with only a small increase in research effort.

9.1. BASIC EVOLUTIONARY PROBLEMS

(1) *The Eusociality Threshold*

Kin selection, reinforced by the haplodiploid bias, appears to have been a principal agent in the origin of eusociality within the order Hymenoptera. But for reasons explained in Chapter One, kin selection cannot be the sole cause. Preadaptations in nest construction, antipredation, parental exploitation of off-spring, and other traits must contribute to increasing inclusive fitness above the "eusociality threshold." Future theory must include an accounting system of all these agents, as well as the equations by which their relative weights can be evaluated.

(2) *What is the Relationship Between Ergonomic Efficiency and Genetic Fitness?*

In the reproductive stage of colony growth, ergonomic efficiency can be translated directly into the production rate of

316

reproductive forms. This rate in turn is a close approximation of genetic fitness when the genetic relatedness of the colony members is taken into account. But this translation is more difficult at earlier stages of the life cycle (Chapters Two, Three). Moreover, no one knows just how to map the natural selection of ergonomic functions onto the genetic variability of the species. As shown in Chapter Eight, this last problem lies at the heart of the application of optimization theory in studies of evolution.

(3) *What are the Relationships Between Processes of*
Natural Selection at the Individual and
Colony Levels?

This complex problem has been explored at some length, especially in Chapter Three, but the theory relating to it remains in a rudimentary state. We suspect that greater progress will come more from imaginative, well-designed experimental studies than from further model building.

9.2. CASTE DIVERSIFICATION

(4) *How Anatomically Different and Anatomically*
Specialized can the Various Castes Become?

The concept of the allometric space (Chapter Five) is a minimal and almost certainly incomplete specification of the variables that have affected the number of castes during colony evolution. Many of these variables, including allometric variation, efficiency curves, "social coverage," and others, should be directly measurable in laboratory and field studies.

(5) *How are Developmental Pathways and Decision*
Points Determined so as to Optimize Caste Ratios?

The growth transformation rules and decision points remain a relatively unexplored area of the biology of social insects (Chapter Four). The few species studied to date display a

317

striking variation in their patterns. The full range of the variation and its ultimate relation to ecological adaptation (if any such relation exists) will perhaps be clarified by an extension of current physiological research on caste.

(6) *How Flexible are Physical Caste Ratios?*

Temporal castes are flexible, in the sense that when faced with emergencies the members of one age cohort can alter their behavior to take up the roles of other age cohorts (Chapter Four). It is not known to what extent the ratios of physical castes are also altered as a response to environmental stress (Chapter Six). This is one of the questions seemingly most tractable to experimental investigation.

9.3. ECOLOGY AND DEMOGRAPHY

(7) *What are the Major Ecological Determinants of Caste Evolution?*

This is the important question explored in Chapters Four to Seven. We considered at some length the possible effects of variation in resource size-frequency distributions, prey abundance, intra- and interspecific competition, and mortality from predation. But this largely deductive exercise was only a small first step. There now exists an obvious need for an entire array of new experimental studies. Although some can be conducted in the laboratory, even greater opportunities exist in the field. The single greatest need appears to us to be more information about the responses of castes to the natural environmental contingencies for which they are presumed to represent adaptations.

(8) *Ontogenetic Changes in Caste Ratios*

As colonies grow in size, caste ratios almost always change, and sometimes to a drastic extent. By relating these changes to the shifts in the influence of various environmental agents on colony welfare, the basic theory of caste evolution can be tested and extended.

(9) *Survivorship Schedules of Individual Castes*

In Chapter Five we postulated a relation between senescence and the varying survivorship schedules of different castes. As a competing hypothesis it was suggested that early senescence is a nonadaptive result of the construction of "cheap" components, which provide greater overall ergonomic efficiency under certain environmental conditions. An empirical examination of these possible relationships could lead to an evaluation of colony-level selection and an improvement of caste theory.

9.4. COMPARATIVE STUDIES

(10) *Ant Versus Termite Caste Systems*

One of the most remarkable of all evolutionary events in the history of life has been the convergence between the caste systems of ants and termites. The ants and termites represent two of the major phylogenetic branches of the insects, separated by over 250 million years of evolution. Also, their way of generating castes is very different: ants diversify individuals within the single adult instar by means of allometric growth in the imaginal discs, whereas termites rely to a large extent on variation among series of instars (Chapter Four). Sociograms of many species will be needed to examine the similarities and differences in detail, in order to evaluate the factors that shaped the convergence.

(11) *The Systems of Ants and Other Hymenopterans*

The weakness of phylogeny as a determinant of caste evolution is further suggested by the dissimilarity of ant castes to those of other social hymenopterans, the bees and wasps. The elaborate physical caste systems of ants and termites may be connected in some way to the flightless condition of the workers and their largely subterranean nest sites, traits that ants share with termites. But if a causal connection exists, it is far from obvious. The difference between ants and other social hymenopterans must be classified as a principal remaining mystery.

(12) *Social Insects Versus the Colonial Invertebrates*

Bryozoans, corals, siphonophores, and other colonial invertebrates surpass the social insects in the diversification and specialization of their castes. This difference is evidently due in part to the asexual reproduction of the colonial invertebrates, which results in complete genetic identity of the zooids. It is apparently also based on the anatomical simplicity and sessile nature of the zooids, which permit them to be physically united. Thus, a colony of invertebrates occupies a grade of evolution intermediate between a colony of social insects and a metazoan organism. An ergonomics of colonial invertebrates is indicated; a beginning has in fact already been made by Schopf (1973). It is possible that common principles will be formed between developmental biology and sociobiology by due attention to ergonomic theory.

Glossary

Alate. Referring to a winged individual (contrast with dealate).

Allometric space. The set of joint measurements in two or more bodily dimensions. Each point in the space represents a particular anatomical type; certain points represent the anatomical types best able to meet particular contingencies in the environment.

Allometry. A size relation between two body parts that can be expressed by $y = bx^a$, where a and b are fitted constants. Physical castes in ants are based on allometry with $a \neq 1$. (See also isometry.)

Altruism. Self-destructive behavior performed for the benefit of others.

Brood. The collective immature members of a colony, including eggs, nymphs, larvae, and pupae.

Caste. Any set of colony members, smaller than the total colony population, that specialize on particular tasks for prolonged periods of time. (See also physical caste and temporal caste.)

Caste distribution function (CDF). The relative frequency of different groups of workers, classified according to size, allometric proportions, age, and tempo of activity.

CDF. See caste distribution function.

Coefficient of relatedness. See degree of relatedness.

Colony. A group of individuals that constructs nests or rears offspring in a cooperative manner.

Dealate. Referring to an individual that has shed its wings, usually after mating; used both as an adjective and a noun (contrast with alate).

Defense distribution function (DDF). The relative frequency of contingencies facing the colony due to attacks by enemies and damage to the nest.

Degree of relatedness (r_{ij}). The fraction of genes that are identical between two individuals due to descent from a common ancestor.

Dimorphism. In caste systems, the existence in the same colony of two different anatomical forms connected by few or no intermediates.

Diphasic allometry. Polymorphism in which the allometric regression line, when plotted on a double logarithmic scale, "breaks" and consists of two segments of different slopes whose ends meet at an intermediate point.

Eclosion. Emergence of the adult from the pupa.

Elitism. The existence of elites, which are colony members displaying much greater than average initiative or activity.

Ergonomics. The quantitative study of work, performance, and efficiency.

Ergonomic efficiency function (η). The task performance efficiency of an individual with particular physical proportions.

Ergonomic stage. The period of growth of a colony, consisting primarily of the addition of workers, that occurs between the founding and reproductive phases.

Eusocial. Referring to "higher" social life in the insects. Specifically, eusocial means the condition or the group possessing it in which individuals display all of the following three traits: cooperation in caring for the young; reproductive division of labor, with more or less sterile individuals working on behalf of individuals engaged in reproduction; and overlap of at least two generations of life stages capable of contributing to colony labor.

FEF. Foraging efficiency function (*q.v.*).

Foraging efficiency function (FEF). The rate and efficiency at which particular castes utilize particular resources; the necessary link between the caste and resource distribution functions.

Gyny. The number of egg-laying queens in a colony (see monogyny and polygyny).

Haplodiploidy. The mode of sex determination in which males are derived from haploid eggs and females from diploid eggs.

Haplometrosis. The founding of a colony by a single egg-laying queen (contrast with pleometrosis).

Hemimetabolous. Undergoing development that is gradual and lacks a sharp separation into larval, pupal, and adult stages. Termites, for example, are hemimetabolous. (Compare with holometabolous.)

Holometabolous. Undergoing a complete metamorphosis during development, with distinct larval, pupal, and adult stages. The Hymenoptera are an example of a holometabolous group of insects. (Compare with hemimetabolous.)

Hymenoptera. The order of insects that includes the ants, bees, and wasps.

Inclusive fitness. The sum of an individual's own genetic fitness plus all its influence on fitness in its relatives other than direct descendants.

Instar. An insect in any one of the growth intervals between molts, or the growth interval itself; the pupa and the adult, for example, each represent a separate instar.

Isometry. Allometry (*q.v.*) in which the exponent $a = 1$, so that large individuals have the same proportions as small ones.

Kin selection. The selection of genes due to one or more individuals favoring or disfavoring the survival and reproduction of relatives who possess the same genes by common descent.

Major worker. A member of the largest worker subcaste, especially in ants. (See also media worker and minor worker.)

Media worker. In polymorphic ant series containing three or more worker subcastes, an individual belonging to the medium-sized subcaste(s). (See also minor worker and major worker.)

Metrosis. The number of egg-laying queens that found a colony (see haplometrosis and pleometrosis).

Minim (or *minima*). A minor worker (*q.v.*), especially one of the smallest members of this subcaste (such individuals are also called nanitics).

Minor worker. A member of the smallest worker subcaste, especially in ants.

Molt. The casting-off of the outgrown skin or exoskeleton in the process of growth.

Monogyny. The existence of only one functional queen in the nest (opposed to polygyny).

Monomorphism. The existence within a species or colony of only a single worker physical subcaste.

Monophasic allometry. Polymorphism in which the allometric regression line has a single slope.

Multicolonial. Pertaining to a population of social insects which is divided into colonies that recognize nest boundaries (opposed to unicolonial).

Nanitic worker. A dwarf worker produced from either the first ant broods or later ant broods that have been subjected to starvation.

Physical caste. A caste distinguished not only by behavior but also by distinctive anatomical traits (opposed to temporal caste).

Pleometrosis. The founding of a colony by two or more egg-laying queens (contrast with haplometrosis).

Polyethism. Division of labor.

Polygyny. The coexistence in the same colony of two or more egg-laying queens (opposed to monogyny).

Polymorphism. In social insects, the coexistence of two or more functionally different physical castes within the same sex. In ants it is possible to define polymorphism more precisely as the occurrence of nonisometric relative growth occurring over a sufficient range of size variation within a normal mature colony to produce individuals of distinctly different proportions at the extremes of the size range.

Queen. A member of the female reproductive caste in eusocial insect species. The existence of a queen caste presupposes the existence also of a worker caste at some stage of the colony life cycle. Queens may or may not be morphologically different from workers.

r_{ij}. The degree of relatedness (*q.v.*).

RDF. Resource distribution function (*q.v.*).

Resource distribution function (RDF). The relative frequency of different classes of food items in the foraging area of a colony, classified according to the amount of time they have been available, their size, and their chemical properties.

Returns to scale. In the present biological context, the rate of increase of productivity as a function of the increase of colony size.

Risk adaptedness factor (ρ). The rate at which the colony fitness, measured by the potential to produce new reproductive forms, increases as a function of the net energetic profit due to foraging. The greater this factor, the more a colony stands to gain (or lose) as a result of fluctuations in the environment.

Social insect. In the strict sense, a "true social insect" is one that belongs to a eusocial species; in other words, it is an ant, a termite, or one of the eusocial wasps or bees.

Soldier. A member of a worker subcaste specialized for colony defense.

Task. A set of behaviors that must be performed to achieve some purpose of the colony, such as repelling an invader or feeding a larva.

Tempo. Activity level.

Temporal caste. A caste distinguished not only by behavior but also by age (contrast with physical caste).

Triphasic allometry. Polymorphism in which the allometric regression line, when plotted on a double logarithmic scale, "breaks" at two points and consists of three segments.

Tychophile species. A species whose caste system and behavior are adapted to absorbing large fluctuations in the environment, either in the food supply or in attacks by predators. Such species gain by the potential for rapid colony growth and reproduction.

Tychophobe species. A species whose caste system and behavior have evolved so as to minimize the effects of environmental fluctuation, but most likely at the cost of a lowered rate of colony growth and reproduction.

Unicolonial. Pertaining to a population of social insects in which there are no behavioral colony boundaries (opposed to multicolonial).

Worker. A member of the nonreproductive, laboring caste in social insects.

Bibliography

Alberch, P., G. F. Oster, S. J. Gould, and D. Wake. 1978. Size and shape in ontogeny and phylogeny. *Journal* of *Paleobiology*, in press.

Alexander, R. D. 1974. The evolution of social behavior. *Annual Review of Ecology and Systematics*, 5:325–383.

Alexander, R. D. and P. W. Sherman. 1977. Local mate competition and parental investment in social insects. *Science*, 196:494–500.

Allen, Elizabeth, et al. (35 authors, comprising the Sociobiology Study Group of Science for the People). 1976. Sociobiology—another biological determinism. *BioScience*, 26(3):182, 184–186.

Aoki, S. 1977. *Colophina clematis* (Homoptera, Pemphigidae), an aphid species with "soldiers." *Kontyû*, Tokyo, 45(2):276–282.

Astrom, K. 1970. *Introduction to Stochastic Control Theory*. Academic Press, New York.

Auslander, D., J. Guckenheimer, and G. F. Oster. 1978. Chaotic evolutionarily stable strategies. *Theoretical Population Biology*, in press.

Barlow, R. and F. Proschan. 1975. *Statistical Theory of Reliability and Life Testing*. Holt, Rinehart and Winston, New York.

Baroni Urbani, C. 1976. Réinterpretation du polymorphisme de la caste ouvrière chez les fourmis à l'aide de la régression polynomiale. *Revue Suisse Zoologique*, 83(1):105–110.

Beig, D. 1972. The production of males in queenright colonies of *Trigona* (*Scaptotrigona*) *postica*. *Journal of Apicultural Research*, 11:33–39.

Benford, F. 1978. Fisher's theory of the sex ratio applied to the social Hymenoptera. *Journal of Theoretical Biology*, 72:701–727.

BIBLIOGRAPHY

Bernstein, Ruth A. 1976. The adaptive value of polymorphism in an alpine ant, *Formica neorufibarbis gelida* Wheeler. *Psyche*, Cambridge, 83(2):181–184.

Bernstein, S. and Ruth A. Bernstein. 1969. Relationship between foraging efficiency and the size of the head and component brain and sensory structures in the red wood ant. *Brain Research*, 16(1):85–104.

Bier, K.-H. 1958. Die Regulation der Sexualität in den Insektenstaaten. *Ergebnisse der Biologie*, 20:97–126.

Bodenheimer, F. S. 1937. Population problems of social insects. *Biological Reviews* (Cambridge Philosophical Society), 12(4):393–430.

Borch, K. 1968. *The Economics of Uncertainty*. Princeton University Press, Princeton, N.J.

Brian, Anne D. 1951. Brood development in *Bombus agrorum* (Hym., Bombidae). *Entomologist's Monthly Magazine*, 87:207–212.

Brian, M. V. 1953. Brood-rearing in relation to worker number in the ant *Myrmica*. *Physiological Zoölogy*, 26(4):355–366.

Brian, M. V. 1955. Studies of caste in *Myrmica rubra* L. (2): The growth of workers and intercastes. *Insectes Sociaux*, 2(1):1–34.

Brian, M. V. 1957a. Caste determination in social insects. *Annual Review of Entomology*, 2:107–120.

Brian, M. V. 1957b. The growth and development of colonies of the ant *Myrmica*. *Insectes Sociaux*, 4(3):177–190.

Brian, M. V. 1965a. *Social Insect Populations*. Academic Press, London.

Brian, M. V. 1965b. Caste differentiation in social insects. Symposium of the Zoological Society of London, 14:13–38.

Brian, M. V. 1968. Regulation of sexual production in an ant society. Colloques Internationaux du Centre National de la Recherche Scientifique, Paris, 1967, No. 173, pp. 61–76.

Brian, M. V. 1973. Feeding and growth in the ant *Myrmica*. *Journal of Animal Ecology*, 42(1):37–53.

Brian, M. V. and Anne D. Brian. 1955. On the two forms macrogyna and microgyna of the ant *Myrmica rubra* L. *Evolution*, 9(3):280–290.

Brothers, D. J. 1975. Phylogeny and classification of the aculeate Hymenoptera, with special reference to the Mutillidae. *Science Bulletin*, University of Kansas, Lawrence, Kansas, 50(11):483–648.

Brothers, D. J. and C. D. Michener. 1974. Interactions in colonies of primitively social bees: III, ethometry of division of labor in *Lasioglossum zephyrum* (Hymenoptera: Halictidae). *Journal of Comparative Physiology*, 90(2):129–168.

Brown, W. L. 1973. A comparison of the Hylean and Congo-West African forest ant faunas. In Betty J. Meggers, E. S. Ayensu, and W. D. Duckworth, eds., *Tropical Forest Ecosystems in Africa and South America: A Comparative Review*, pp. 161–185. Smithsonian Institution Press, Washington, D.C.

Brown, W. L. and E. O. Wilson. 1959. The evolution of the dacetine ants. *Quarterly Review of Biology*, 34:278–294.

Bryson, A. and Y. Ho. 1975. *Applied Optimal Control*. Wiley, New York.

Buschinger, A. 1974. Monogynie und Polygynie in Insektensozietäten. In G. H. Schmidt, ed. (*q.v.*), *Sozialpolymorphismus bei Insekten*, pp. 862–896.

Buschinger, A. 1975. Eine genetische Komponente im Polymorphismus der dulotische Ameise *Harpagoxenus sublaevis*. *Naturwissenschaften*, 62(5):239.

Cammaerts-Tricot, M.-C. 1974. Production and perception of attractive pheromones by differently aged workers of *Myrmica rubra* (Hymenoptera: Formicidae). *Insectes Sociaux*, 21(3):235–247.

Chadab, Ruth and C. W. Rettenmeyer. 1975. Mass recruitment by army ants. *Science*, 188:1124–1125.

Chen, S. C. 1937. The leaders and followers among the ants in nest-building. *Physiological Zoölogy*, 10(4):437–455.

Cherrett, J. M. 1972. Some factors involved in the selection of vegetable substrate by *Atta cephalotes* (L.) (Hymenoptera: Formicidae) in tropical rain forest. *Journal of Animal Ecology*, 41:647–660.

Cohen, D. and I. Eshel. 1976. On the founder effect and the evolution of altruistic traits. *Theoretical Population Biology*, 10:276–302.

Cohen, J. E. 1971. Mathematics as metaphor. [Review of *Dynamical Systems Theory in Biology*, Vol. 1, by R. Rosen; 1970.] *Science*, 172:674–675.

Cohen, J. E. 1976. Irreproducible results and the breeding of pigs (or nondegenerate limit random variables in biology). *BioScience*, 26(6):391–394.

Corn, M. Lynne. 1976. The ecology and behavior of *Cephalotes atratus*, a Neotropical ant (Hymenoptera: Formicidae). Ph.D. thesis, Harvard University, Cambridge, Mass.

Crozier, R. H. 1974. Allozyme analysis of reproductive strategy in the ant *Aphaenogaster rudis*. *Isozyme Bulletin*, 7:18.

Cumber, R. A. 1949. The biology of humble-bees, with special reference to the production of the worker caste. *Transactions of the Royal Entomological Society of London*, 100(1):1–45.

Darchen, R. and Bernadette Delage-Darchen. 1974. Nouvelles expériences concernant le déterminisme des castes chez les Mélipones (Hyménoptères apidés). *Compte Rendu de l'Académie des Sciences*, Paris, 278:907–910.

Davidson, Diane W. 1978. Size variability in the worker caste of a social insect (*Veromessor pergandei* Mayr) as a function of the competitive environment. *American Naturalist*, 112:523–532.

Delage-Darchen, Bernadette. 1972. Le polymorphisme chez les fourmis *Nematocrema* d'Afrique. *Insectes Sociaux*, 19(3):259–278.

Denholm, J. 1975. Necessary condition for maximum yield in a senescing two-phase plant. *Journal of Theoretical Biology*, 52:251–254.

BIBLIOGRAPHY

Desoer, C. 1970. *Notes for a Second Course in Systems Theory*. Van Nostrand, New York.

Dingle, H. 1972. Aggressive behavior in stomatopods and the use of information theory in the analysis of animal communication. In H. E. Winn and B. L. Ollâ, eds., *Behavior of Marine Animals: Current Perspectives in Research, Vol. 1, Invertebrates*, pp. 126–156. Plenum Press, New York.

Dirac, P.A.M. 1963. The evolution of the physicist's picture of nature. *Scientific American*, 208(5) (May): 45–53.

Dlussky, G. 1975. Formicidae. In A. P. Rasnitsyn, ed., *Hymenoptera Apocrita of Mesozoic*. Transactions of the Paleontological Institute, Academy of Sciences of the USSR, 147:114–122.

Dobrzańska, Janina. 1966. The control of territory by *Lasius fuliginosus* Latr. *Acta Biologiae Experimentalis*, Warsaw, 26(2):193–213.

Dorny, C. 1975. *A Vector Space Approach to Models and Optimization*. Wiley, New York.

Eberhard, Mary Jane West. 1969. The social biology of polistine wasps. *Miscellaneous Publications*, Museum of Zoology, University of Michigan, 140:1–101.

Eisner, T. 1970. Chemical defense against predation in arthropods. In E. Sondheimer and J. B. Simeone, eds., *Chemical Ecology*, pp. 157–217. Academic Press, New York.

Eisner, T., Irmgard Kriston, and D. J. Aneshansley. 1976. Defensive behavior of a termite (*Nasutitermes exitiosus*). *Behavioral Ecology and Sociobiology*, 1(1):83–125.

Ekeland, I. 1977. *Discontinuité des champs Hamiltonien et existance de solutions optimales en calcul de variations*. Cahiers de Mathematiques de la Decision, in press.

Elmes, G. W. 1973. Observations on the density of queens in natural colonies of *Myrmica rubra* L. (Hymenoptera: Formicidae). *Journal of Animal Ecology*, 42(3):761–771.

Evans, H. E. and Mary Jane West Eberhard. 1970. *The Wasps*. University of Michigan Press, Ann Arbor.

331

Fagen, R. M. and R. Goldman. 1977. Behavioural catalogue analysis methods. *Animal Behaviour*, 25:261–274.

Friedman, M. 1953. *Essays in Positive Economics*. University of Chicago Press, Chicago.

Frisch, K. von. 1967. *The Dance Language and Orientation of Bees* (tr. L. Chadwick). Belknap Press of Harvard University Press, Cambridge, Mass.

Goll, W. 1967. Strukturuntersuchungen am Gehirn von *Formica*. *Zeitschrift für Morphologie und Ökologie der Tiere*, 59(2):143–210.

Golley, F. B. and J. B. Gentry. 1964. Bioenergetics of the southern harvester ant, *Pogonomyrmex badius*. *Ecology*, 45(2):217–225.

Gottfried, B. and J. Weisman. 1973. *Introduction to Optimization Theory*. Prentice-Hall, Englewood Cliffs, N.J.

Gotwald, W. H. 1971. Phylogenetic affinity of the ant genus *Cheliomyrmex* (Hymenoptera: Formicidae). *Journal of the New York Entomological Society*, 79(3):161–173.

Gould, S. J. 1966. Allometry and size in ontogeny and phylogeny. *Biological Reviews* (Cambridge Philosophical Society), 41:587–640.

Gregg, R. E. 1942. The origin of castes in ants with special reference to *Pheidole morrisi* Forel. *Ecology*, 23(3):295–308.

Guckenheimer, J., G. F. Oster, and A. Ipaktchi. 1977. The dynamics of density dependent population models. *Journal of Mathematical Biology*, 4:101–147.

Hamilton, W. D. 1964. The genetical theory of social behaviour, I, II. *Journal of Theoretical Biology*, 7(1):1–52.

Hamilton, W. D. 1967. Extraordinary sex ratios. *Science*, 156:477–488.

Hamilton, W. D. 1972. Altruism and related phenomena, mainly in social insects. *Annual Review of Ecology and Systematics*, 3:193–232.

Hamilton, W. J., III. 1973. *Life's Color Code*. McGraw-Hill, New York.

Haverty, M. I. 1977. The proportion of soldiers in termite colonies: a list and a bibliography. *Sociobiology* (Chico, California), 2(3):199–217.

Haverty, M. I., W. L. Nutting, and J. P. LaFage. 1978. Caste composition and soldier proportions in five species of southern Arizona termites. Manuscript in preparation.

Hecker, H. 1966. Das Zentralnervensystem des Kopfes und seine postembryonale Entwicklung bei *Bellicositermes bellicosus* (Smeath.) (Isoptera). *Acta Tropica*, 23(4):297–352.

Heinrich, B. 1978. *Bumblebee Economics*. Harvard University Press, Cambridge, Mass., in press.

Hemmingsen, A. M. 1973. Nocturnal weaving on nest surface and division of labour in weaver ants (*Oecophylla smaragdina* Fabricius, 1775). *Videnskabelige Meddelelser fra Dansk Naturhistorisk Forening*, 136:49–56.

Herbers, Joan M. 1977. Behavioral constancy in *Formica obscuripes* (Hymenoptera: Formicidae). *Annals of the Entomological Society of America*, 70(4):485–486.

Hermann, H. R. 1975. The ant-like venom apparatus of *Typhoctes peculiaris*, a primitive mutillid wasp. *Annals of the Entomological Society of America*, 68(5):882–884.

Higashi, S. 1974. Worker polyethism related with body size in a polydomous red wood ant, *Formica (Formica) yessensis* Forel. *Journal of the Faculty of Science*, Hokkaido University, series 6, 19(3):695–705.

Hölldobler, B. 1962. Zur Frage der Oligogynie bei *Camponotus ligniperda* Latr. und *Camponotus herculeanus* L. (Hym. Formicidae). *Zeitschrift für Angewandte Entomologie*, 49(4):337–352.

Hölldobler, B. 1966. Futterverteilung durch Männchen in Ameisenstaat. *Zeitschrift für Vergleichende Physiologie*, 52(4):430–455.

Hölldobler, B. 1973. Chemische Strategie beim Nahrungserwerb der Diebsameise (*Solenopsis fugas* Latr.) und der

Pharaoameise (*Monomorium pharaonis* L.). *Oecologia*, Berlin, 11:371–380.

Hölldobler, B. 1976. Recruitment behavior, home range orientation and territoriality in harvester ants, *Pogonomyrmex*. *Behavioral Ecology and Sociobiology*, 1:3–44.

Hölldobler, B. 1977. Communication in social Hymenoptera. In T. Sebeok, ed., *How Animals Communicate*, pp. 418–471. Indiana University Press, Bloomington, Indiana.

Hölldobler, B. and E. O. Wilson. 1977a. The number of queens: an important trait in ant evolution *Naturwissenschaften*, 64:8–15.

Hölldobler, B. and E. O. Wilson. 1977b. Weaver ants: social establishment and maintenance of territories. *Science*, 195:900–902.

Holmes, P. and D. Rand. 1977. Bifurcations of the forced van der Pol oscillator. Unpublished manuscript.

Holt, S. J. 1955. On the foraging activity of the wood ant. *Journal of Animal Ecology*, 24(1):1–34.

Horstmann, K. 1973. Untersuchungen zur Arbeitsteilung unter den Aussendienst-arbeiterinnen der Waldameise *Formica polyctena* Foerster. *Zeitschrift für Tierpsychologie*, 32(5):532–543.

Howard, R. 1971. *Dynamic Probabilistic Systems, Vol. 1, Markov Models; Vol. 2, Semi-Markov and Decision Models*. Wiley, New York.

Howse, P. E. 1968. On the division of labour in the primitive termite *Zootermopsis nevadensis* (Hagen). *Insectes Sociaux*, 15(1):45–50.

Hrdý, I. 1972. Der Einfluss von zwei Juvenilhormonanalogen auf die Differenzierung der Soldaten bei *Reticulitermes lucifugus santonensis* Feyt (Isopt.: Rhinotermitidae). *Zeitschrift für Angewandte Entomologie*, 72(2):129–134.

Huxley, J. 1932. *Problems of Relative Growth*. Dial Press, New York.

Intriligator, M. 1971. *Mathematical Optimization and Economic Theory*. Prentice-Hall, Englewood Cliffs, N.J.

Ishay, J., H. Bytinsky-Salz, and A. Shulov. 1967. Contributions to the bionomics of the Oriental hornet (*Vespa orientalis* Fab.). *Israel Journal of Entomology*, 2:45–106.

Ishay, J. and R. Ikan. 1969. Gluconeogenesis in the Oriental hornet *Vespa orientalis* F. *Ecology*, 49(1):169–171.

Jacob, F. 1977. Evolution and tinkering. *Science*, 196:1161–1166.

Jaisson, P. 1975. L'impregnation dans l'ontogenese des comportements de soins aux cocons chez la jeune fourmi rousse (*Formica polyctena* Forst.). *Behaviour*, 52(1,2):1–37.

Jander, R. 1957. Die optische Richtungsorienterung der Roten Waldameise (*Formica rufa* L.). *Zeitschrift für Vergleichende Physiologie*, 40(2):162–238.

Janzen, D. H. 1973a. Evolution of polygynous obligate acacia-ants in western Mexico. *Journal of Animal Ecology*, 42:727–750.

Janzen, D. H. 1973b. Sweep samples of tropical foliage insects: description of study sites, with data on species abundances and size distributions. *Ecology*, 54(3):659–686.

Janzen, D. H. and T. W. Schoener. 1968. Differences in insect abundance and diversity between wetter and drier sites during a tropical dry season. *Ecology*, 49(1):96–110.

Jeanne, R. L. 1972. Social biology of the Neotropical wasp *Mischocyttarus drewseni*. *Bulletin of the Museum of Comparative Zoology*, Harvard University, Cambridge, Mass., 144(3):63–150.

Johnson, N. and S. Kotz. 1970. *Continuous Univariate Distributions-2*. Houghton-Mifflin, Boston.

Kerr, W. E. 1950. Genetic determination of castes in the genus *Melipona*. *Genetics*, 35(2):143–152.

Kerr, W. E. 1962. Genetics of sex determination. *Annual Review of Entomology*, 7:157–176.

335

Kerr, W. E. 1974. Geschlechts- und Kastendetermination bei stachellosen Bienen. In G. H. Schmidt, ed. (*q.v.*), *Sozialpolymorphismus bei Insekten*, pp. 336–349.

Kerr, W. E. and N. J. Hebling. 1964. Influence of the weight of worker bees on division of labor. *Evolution*, 18(2):267–270.

Kerr, W. E. and W. E. Nielsen. 1966. Evidences that genetically determined *Melipona* queens can become workers. *Genetics*, 54(3):859–866.

Kleinrock, L. 1976. *Queueing Systems*, Vol. 1. Wiley, New York.

Kratky, E. 1931. Morphologie und Physiologie der Drüsen im Kopf und Thorax der Honigbiene (*Apis mellifica* L.). *Zeitschrift für Wissenschaftliche Zoologie*, 139:120–200.

Krebs, C. J., M. S. Gaines, B. L. Keller, Judith H. Myers, and R. H. Tamarin. 1973. Population cycles in small rodents. *Science*, 179:35–44.

Leitmann, G. 1966. *An Introduction to Optimal Control*. McGraw-Hill, New York.

Lévieux, J. 1972. Le role des fourmis dan les réseaux trophiques d'une savane préforestière de Côte-d'Ivoire. *Annales de l'Université d'Abidjan*, series E, 5(1):143–240.

Levin, D. and M. Kerster. 1969. Density-dependent gene dispersal in *Liatris*. *American Naturalist*, 103:61–73.

Levin, S. and J. D. Udovic. 1977. A mathematical model of coevolving populations. *American Naturalist*, 111:657–675.

Levins, R. 1968. *Evolution in Changing Environments*. Princeton University Press, Princeton, N.J.

Levins, R. 1977. The limits of optimization. Unpublished manuscript.

Lewis, T., G. V. Pollard, and G. C. Dibley. 1974. Rhythmic foraging in the leaf-cutting ant *Atta cephalotes* (L.) (Formicidae: Attini). *Journal of Animal Ecology*, 43:129–141.

Lewontin, R. C. 1977a. Adaptation. In *The Italian Encyclopedia*. Einaudi, Turin.

Lewontin, R. C. 1977b. Fitness, survival and optimality. In D. J. Horn, ed., *Ohio State University Colloquium on Ecology*. Ohio State University Press, Columbus.

Lin, N. and C. D. Michener. 1972. Evolution of sociality in insects. *Quarterly Review of Biology*, 47:131–159.

Lindauer, M. 1961. *Communication Among Social Bees*. Harvard University Press, Cambridge, Mass.

Lüscher, M. 1974. Kasten und Kastendifferenzierung bei niederen Termiten. In G. H. Schmidt, ed. (*q.v.*), *Sozialpolymorphismus bei Insekten*, pp. 694–739.

Lüscher, M., ed. 1977. *Phase and Caste Determination in Insects: Endocrine Aspects* (Fifteenth International Congress of Entomology, Washington, D.C., August, 1976). Pergamon Press, Elmsford, N.Y.

Macevicz, S. 1978. Some consequences of Fisher's sex ratio principle for social Hymenoptera that reproduce by colony fission. *American Naturalist*, in press.

Macevicz, S. and G. F. Oster. 1976. Modeling social insect populations II: optimal reproductive strategies in annual eusocial insect colonies. *Behavioral Ecology and Sociobiology*, 1(3):265–282.

Markowitz, H. 1959. *Portfolio Selection*. Wiley, New York.

Maschwitz, U. 1966. Das Speichelsekret der Wespenlarven und seine biologische Bedeutung. *Zeitschrift für Vergleichende Physiologie*, 53(3):228–252.

Maschwitz, U. and E. Maschwitz. 1974. Platzende Arbeiterinnen: Eine neue Art der Feindabwehr bei sozialen Hautflüglern. *Oecologia* (Berlin), 14(3):289–294.

May, R. M. 1973. *Stability and Complexity in Model Ecosystems*. Princeton University Press, Princeton, N.J.

May, R. M. 1976. Simple mathematical models with very complicated dynamics. *Nature*, 261:459–467.

May, R. M. and G. F. Oster. 1976. Bifurcations and dynamic complexity in simple ecological models. *American Naturalist*, 110:573–599.

Maynard Smith, J. 1974. The theory of games and the evolution of animal conflicts. *Journal of Theoretical Biology*, 47:209–221.

Maynard Smith, J. 1976. Evolution and the theory of games. *American Scientist*, 64(1):41–45.

Medawar, P. B. 1957. *The Uniqueness of the Individual*. Methuen, London.

Melsa, J. and A. Sage. 1973. *An Introduction to Probability and Stochastic Processes*. Prentice-Hall, Englewood Cliffs, N.J.

Meudec, M. 1973. Sur les variations temporelles du comportement de transport du couvain dans un lot d'ouvrières de *Tapinoma erraticum* Latr. (Formicidae Dolichoderinae). *Compte Rendu de l'Académie des Sciences*, Paris, 208(11):831–832.

Michener, C. D. 1964. Reproductive efficiency in relation to colony size in hymenopterous societies. *Insectes Sociaux*, 11(4):317–341.

Michener, C. D. 1974. *The Social Behavior of the Bees: A Comparative Study*. Belknap Press of Harvard University Press, Cambridge, Mass.

Michener, C. D. and D. J. Brothers. 1974. Were workers of eusocial Hymenoptera initially altruistic or oppressed? *Proceedings of the National Academy of Sciences*, USA, 71:671–674.

Miquel, J. 1971. Aging of male *Drosophila melanogaster*: histological, histochemical, and ultrastructural observations. *Advances in Gerontological Research*, 3:39–71.

Mirmirani, M. and G. F. Oster. 1978. Competition, kin selection and evolutionarily stable strategies. *Theoretical Population Biology*, in press.

Möglich, M. and B. Hölldobler. 1974. Social carrying behavior and division of labor during nest moving in ants. *Psyche*, Cambridge, 81(2):219–236.

Möglich, M. and B. Hölldobler. 1975. Communication and

orientation during foraging and emigration in the ant *Formica fusca. Journal of Comparative Physiology*, 101(4):275–288.

Montagner, H. 1963. Étude préliminaire des relations entre les adultes et le couvain chez les guêpes sociales du genre *Vespa*, au moyen d'un radio-isotope. *Insectes Sociaux*, 10(2):153–165.

Montagner, H. 1966. Sur l'origine des mâles dans les societées de guêpes du genre *Vespa. Compte Rendu de l'Académie Sciences*, Paris, 263:785–787.

Moore, P. 1954. Spacing in plant populations. *Ecology*, 35:222–227.

Mossin, J. 1968. Optimal multiperiod portfolio policies. *Journal of Business*, 41:215–229.

Noirot, C. 1969. Formation of castes in the higher termites. In K. Krishna and Frances M. Weesner, eds., *Biology of Termites*, Vol. 1, pp. 311–350. Academic Press, New York.

Nolan, W. J. 1924. The division of labor in the honeybee. *North Carolina Beekeeper*, (October): 10–15.

Nolte, D. J., I. Dési, and Beryl Meyers. 1969. Genetic and environmental factors affecting chiasma formation in locusts. *Chromosoma*, Berlin, 27(2):145–155.

Oster, G. F. 1976. Modeling social insect populations. I. Ergonomics of foraging and population growth in bumblebees. *American Naturalist*, 110:215–245.

Oster, G. F. and J. Guckenheimer. 1977. Bifurcation behavior of population models. In J. Marsden and M. McCracken, eds., *The Hopf Bifurcation, Lecture Notes in Mathematics*, Vol. 19, pp. 327–353. Springer-Verlag, New York.

Oster, G. F., I. Eshel, and D. Cohen. 1977. Worker-queen conflict and the evolution of social castes. *Theoretical Population Biology*, 12:49–85.

Oster, G. F., A. Ipaktchi, and S. Rocklin. 1976. Phenotypic

structure and bifurcation behavior of population models. *Theoretical Population Biology*, 10:365–382.

Oster, G. F. and S. Rocklin. 1978. Optimization models in evolutionary biology. In S. Levin ed., *Some Mathematical Questions in Biology*. American Mathematical Society, Providence, Rhode Island.

Otto, D. 1958. Über die Arbeitsteilung im Staate von *Formica rufa rufo-pratensis minor* Gössw. und ihre verhaltensphysiologischen Grundlagen. *Wissenschaftliche Abhandlungen der Deutschen Akademie der Landwirtschaftswissenschaften zu Berlin*, 30:1–169.

Owen, G. 1968. *Game Theory*. Saunders, Philadelphia.

Paloheimo, J. 1971a. A stochastic theory of search: implications for predator-prey systems. *Mathematical Biology*, 12:105–132.

Paloheimo, J. 1971b. On a theory of search. *Biometrika*, 58:61–75.

Passera, L. 1974. Différenciation des soldats chez la fourmi *Pheidole pallidula* Nyl. (Formicidae Myrmicinae). *Insectes Sociaux*, 21(1):71–86.

Pasteels, J. M. 1965. Polyéthisme chez les ouvriers de *Nasutitermes lujae* (Termitidae Isoptères). *Biologica Gabonica*, 1(2):191–205.

Perelson, A., M. Mirmirani, and G. F. Oster. 1976. Optimal strategies in immunology I: B-cell differentiation and proliferation. *Journal of Mathematical Biology*, 3:325–367.

Perelson, A., M. Mirmirani, and G. F. Oster. 1978. Optimal strategies in immunology II: Memory cell production. *Journal of Mathematical Biology*, in press.

Pielou, E. C. 1969. *An Introduction to Mathematical Ecology*. Wiley, New York.

Pimentel, D. 1968. Population regulation and genetic feedback. *Science*, 159:1432–1437.

Platt, J. R. 1964. Strong inference. *Science*, 146:347–353.

Plowright, R. C. and S. C. Jay. 1968. Caste differentiation in bumblebees (*Bombus* Latr.: Hym.). I. The determination of female size. *Insectes Sociaux*, 15(2):171–192.

Plowright, R. C. and S. C. Jay. 1978. On the size determination of bumblebee castes. *Canadian Journal of Zoology*, in press.

Rapport, D. and J. Turner. 1977. Economic models in ecology. *Science*, 195:367–373.

Rembold, H. 1974. Die Kastenbildung bei der Honigbiene *Apis mellifica* L., aus biochemischer Sicht. In G. H. Schmidt, ed. (*q.v.*), *Sozialpolymorphismus bei Insekten*, pp. 350–403.

Renoux, J. 1976. Le polymorphisme de *Schedorhinotermes lamanianus* (Sjöstedt) (Isoptera: Rhinotermitidae). *Insectes Sociaux*, 23(3):279–494.

Richards, O. W. and Maud J. Richards. 1951. Observations on the social wasps of South America (Hymenoptera: Vespidae). *Transactions of the Royal Entomological Society of London*, 102(1):1–170.

Rocklin, S. and G. F. Oster. 1976. Competition between phenotypes. *Journal of Mathematical Biology*, 3:225–261.

Rockstein, M. 1950. Longevity in the adult worker honeybee. *Annals of the Entomological Society of America*, 43(1):152–154.

Rösch, G. A. 1930. Untersuchungen über die Arbeitsteilung im Bienenstaat. 2. Teil. Die Tätigkeiten der Arbeitsbienen unter experimentell veränderten Bedingungen. *Zeitschrift für Vergleichende Physiologie*, 12(1):1–71.

Röseler, P.-F. 1974. Grossenpolymorphismus, Geslechtsregulation und Stabilisierung der Kasten im Hummelvolk. In G. H. Schmidt, ed. (*q.v.*), *Sozialpolymorphismus bei Insekten*, pp. 298–335.

Rosengren, R. 1971. Route fidelity, visual memory and recruitment behaviour in foraging wood ants of the genus *Formica* (Hymenoptera, Formicidae). *Acta Zoologica Fennica*, 133:1–105.

Roughgarden, J. 1978. *Theory of Population Genetics and Evolutionary Ecology, an Introduction*. Macmillan, New York.

341

BIBLIOGRAPHY

Sahlins, M. 1976. *The Use and Abuse of Biology*. University of Michigan Press, Ann Arbor.

Sakagami, S. F. and H. Fukuda. 1968. Life tables for worker honeybees. *Researches on Population Ecology*, 10(2):127–139.

Schmidt, G. H., ed. 1974. *Sozialpolymorphismus bei Insekten: Probleme der Kastenbildung im Tierreich*. Wissenschaftliche Verlagsgesellschaft MBH, Stuttgart.

Schmidt, G. H. 1974. Steuerung der Kastenbildung und Geschlechtsregulation im Waldameisenstaat. In G. H. Schmidt, ed. (*q.v.*), *Sozialpolymorphismus bei Insekten*, pp. 404–512.

Schneirla, T. C. 1971. *Army Ants: A Study in Social Organization*, ed. H. R. Topoff. W. H. Freeman, San Francisco.

Schneirla, T. C., R. R. Gianutsos, and B. Pasternack. 1968. Comparative allometry in the larval broods of three army-ant genera, and differential growth as related to colony behavior. *American Naturalist*, 102:533–554.

Schoener, T. W. and D. H. Janzen. 1968. Notes on environmental determinants of tropical versus temperate insect size patterns. *American Naturalist*, 102:207–224.

Schopf, T. J. M. 1973. Ergonomics of polymorphism: its relation to the colony as the unit of natural selection in species of the phylum Ectoprocta. In R. S. Boardman, A. H. Cheetham, and W. A. Oliver, Jr., eds., *Animal Colonies: Development and Function through Time*, pp. 247–294. Dowden, Hutchinson, and Ross, Stroudsbury, Pa.

Shashahani, S. 1978. A new mathematical framework for the study of linkage and selection. *Bulletin of the American Mathematical Society*, in press.

Shelton, J. 1960. Solution methods for waiting line problems. *Journal of Industrial Engineering*, 9:293–303.

Silva, Maria M. T. G. da. 1972. Contribução ao estudo da biologia de *Eciton burchelli* Westwood (Hymenoptera, Formicidae). S.D. (Doutor em Ciências) thesis, University of São Paulo, Brazil.

Skellam, J. G., M. V. Brian, and J. R. Proctor. 1959. The simultaneous growth of interacting systems. *Acta Biotheoretica*, 13(2–3):131–144.

Slatkin, M. 1978. On the equilibrium of fitness by natural selection. Unpublished manuscript.

Spradbery, J. P. 1965. The social organization of wasp communities. *Symposium of the Zoological Society of London*, 14:61–96.

Spradbery, J. P. 1973. *Wasps: An Account of the Biology and Natural History of Solitary and Social Wasps*. Sidgwick and Jackson, London.

Tafuri, J. F. 1955. Growth and polymorphism in the larva of the army ant *Eciton (E.) hamatum* Fabricius. *Journal of the New York Entomological Society*, 63:21–40.

Taylor, F. W. 1977. Foraging behavior of ants: experiments with two species of myrmicine ants. *Behavioral Ecology and Sociobiology*, 2(2):147–167.

Taylor, F. W. 1978. Foraging behavior in ants: theoretical considerations. *Journal of Theoretical Biology*, 71:541–563.

Topoff, H. 1971. Polymorphism in army ants related to division of labor and colony cyclic behavior. *American Naturalist*, 105:529–548.

Topoff, H., Katherine Lawson, and Patricia Richards. 1972. Trail following and its development in the Neotropical army ant genus *Eciton* (Hymenoptera: Formicidae: Dorylinae). *Psyche*, Cambridge, 79(4):357–364.

Trivers, R. L. and Hope Hare. 1976. Haplodiploidy and the evolution of the social insects. *Science*, 191:249–263.

Ullmann, J. 1976. *Quantitative Methods in Management*. McGraw-Hill, New York.

Vajda, S. 1972. *Probabilistic Programming*. Academic Press, New York.

Varaiya, P. 1973. *Notes on Optimization*. Van Nostrand Reinhold, New York.

Waloff, N. 1957. The effect of the number of queens of the ant *Lasius flavus* (Fab.) (Hym., Formicidae) on their survival

343

and on the rate of development of the first brood. *Insectes Sociaux*, 4(4):391–408.

Watson, J. A. L. 1974. The development of soldiers in incipient colonies of *Mastotermes darwiniensis* Froggatt (Isoptera). *Insectes Sociaux*, 21(2):181–190.

Weaver, N. 1957. Effects of larval age on dimorphic differentiation of the female honey bee. *Annals of the Entomological Society of America*, 50(3):283–294.

Weber, N. A. 1972. *Gardening Ants: The Attines*. American Philosophical Society, Philadelphia.

Wehner, R. 1969. Die optische Orientierung nach Schwarz-Weiz-Mustern bei verschiedenen Grössenklassen von *Cataglyphis bicolor* Fab. (Formicidae, Hymenoptera). *Revue Suisse Zoologique*, 76(13):371–381.

Went, F. W., Jeanette Wheeler, and G. C. Wheeler. 1972. Feeding and digestion in some ants (*Veromessor* and *Manica*). *BioScience*, 22(2):82–88.

West Eberhard, Mary Jane. 1975. The evolution of social behavior by kin selection. *Quarterly Review of Biology*, 50:1–33.

Wheeler, W. M. 1918. A study of some ant larvae with a consideration of the origin and meaning of social habits among insects. *Proceedings of the American Philosophical Society*, 57:293–343.

Williams, G. C. 1957. Pleiotropy, natural selection, and the evolution of senescence. *Evolution*, 11(4):398–411.

Wilson, D. S. 1975. A theory of group selection. *Proceedings of the National Academy of Sciences*, USA, 72(1):143–146.

Wilson, E. O. 1953. The origin and evolution of polymorphism in ants. *Quarterly Review of Biology*, 28(2):136–156.

Wilson, E. O. 1963. Social modifications related to rareness in ant species. *Evolution*, 17(2):249–253.

Wilson, E. O. 1966. Behaviour of social insects. *Symposium of the Royal Entomological Society of London*, 3:81–96.

Wilson, E. O. 1968. The ergonomics of caste in the social insects. *American Naturalist*, 102:41–66.

Wilson, E. O. 1971. *The Insect Societies.* Belknap Press of Harvard University Press, Cambridge, Mass.

Wilson, E. O. 1972. Animal communication. *Scientific American,* 227(3) (September): 52–60.

Wilson, E. O. 1974a. The population consequences of polygyny in the ant *Leptothorax curvispinosus* Mayr (Hymenoptera: Formicidae). *Annals of the Entomological Society of America,* 67(5): 781–786.

Wilson, E. O. 1974b. Aversive behavior and competition within colonies of the ant *Leptothorax curvispinosus* Mayr (Hymenoptera: Formicidae). *Annals of the Entomological Society of America,* 67(5): 777–780.

Wilson, E. O. 1975a. *Sociobiology: The New Synthesis.* Belknap Press of Harvard University Press, Cambridge, Mass.

Wilson, E. O. 1975b. *Leptothorax duloticus* and the beginnings of slavery in ants. *Evolution,* 29(1): 108–119.

Wilson, E. O. 1975c. Enemy specification in the alarm-recruitment system of an ant. *Science,* 190: 798–800.

Wilson, E. O. 1976a. Behavioral discretization and the number of castes in an ant species. *Behavioral Ecology and Sociobiology,* 1(2): 141–154.

Wilson, E. O. 1976b. A social ethogram of the Neotropical arboreal ant *Zacryptocerus varians* (Fr. Smith). *Animal Behaviour,* 24(2): 354–363.

Wilson, E. O. 1976c. Some central problems of sociobiology. *Social Sciences Information, Biology and Social Life,* 14(6): 5–18. Also, in R. M. May, ed., 1976. *Theoretical Ecology,* pp. 205–217. Saunders, Philadelphia.

Wilson, E. O. 1977. Animal and human sociobiology. *Proceedings of the Academy of Natural Sciences, Philadelphia,* Special Publication No. 12, pp. 273–281.

Wilson, E. O. 1978. Division of labor based on physical castes in fire ants (Hymenoptera: Formicidae: *Solenopsis*). *Journal of the Kansas Entomological Society,* in press.

Wilson, E. O. and B. Hölldobler. 1978. Cooperative ant larvae:

a new test of kin selection theory. Manuscript in preparation.

Wilson, E. O. and G. L. Hunt. 1966. Habitat selection by the queens of two field-dwelling species of ants. *Ecology*, 47(3):485–487.

Wossy, C. 1978. Reproductive success in Hymenoptera. *Theoretical Population Biology*, in press.

Wüst, Margarete. 1973. Stomodeal und proctodeale Sekrete von Ameisenlarven und ihre biologische Bedeutung. *Proceedings of the Seventh Congress of the International Union for the Study of Social Insects*, London, pp. 412–417.

Yamanaka, M. 1928. On the male of a paper wasp, *Polistes fadwigae* Dalla Torre. *Science Reports of the Tôhoku Imperial University*, Sendai, Japan, series 4 (Biol.), 3(3):265–269.

Index

347

349

351

Library of Congress Cataloging in Publication Data

Oster, George F., 1940–
    Caste and ecology in the social insects.

    (Monographs in population biology; 12)
    Bibliography: p.
    1. Insect societies.   2. Insects—Ecology.
I. Wilson, Edward Osborne, 1929– joint author.
II. Title.   III. Series.
QL496.O77   595.7'05'24   78-51185
ISBN 0-691-08210-3
ISBN 0-691-08213-8 pbk.